ANCIENT CULTURE AND SOCIETY

EARLY GREEK SCIENCE
Thales to Aristotle

ANCIENT CULTURE AND SOCIETY

General Editor
M. I. FINLEY
*Professor of Ancient History
in the University of Cambridge*

EARLY
GREEK SCIENCE
Thales to Aristotle

G. E. R. LLOYD

W · W · NORTON & COMPANY

New York · London

W. W. Norton & Company, Inc., 500 Fifth Avenue, New York, N.Y. 10110
www.wwnorton.com

W. W. Norton & Company Ltd., Castle House, 75/76 Wells Street, London W1T 3QT

ISBN 0-393-00583-6
ISBN 978-0-393-00583-7

6 7 8 9 0

CONTENTS

DIAGRAMS

MAP

Note on Pronunciation

In Greek the letter *e* is always a sign of a new syllable, unlike the *e* in *bone*. It is either short, as in *met*, or long, in which case it is pronounced as the *ê* in the French *tête*. In the Greek names and words mentioned in this book the vowel is short unless marked long in the index.

Acknowledgement

The author and publishers are grateful to the Clarendon Press for permission to quote from the Oxford translation, *The Works of Aristotle translated into English*, edited by W. D. Ross.

CHRONOLOGICAL TABLE

The births and deaths of most of the important thinkers in the history of science in our period cannot be determined precisely. Except in the cases of Plato and Aristotle, the dates in the left-hand column are intended merely as a rough guide to the *'floruit'* of the individual concerned, that is, the period at which he may be presumed to have done his chief work.

Scientists			*Contemporary Events*
		c. 610	Thrasybulus, tyrant of Miletus
		594	Solon's archonship
Thales of Miletus	585		
Anaximander of Miletus	555		
		c. 545	Pisistratus in power at Athens
Anaximenes of Miletus	535		
Pythagoras of Samos	525		
		c. 523	Death of Polycrates of Samos
Xenophanes of Colophon	520		
		510	War of Sybaris and Croton
		508	Cleisthenes' reforms
Heraclitus of Ephesus	500		
		494	Miletus destroyed
		490	Battle of Marathon
Parmenides of Elea	480		
		478	Delian League formed
Alcmaeon of Croton	450		
Zeno of Elea			
Anaximander of Clazomenae	445		
Empedocles of Acragas			
Melissus of Samos	440		
Leucippus of Miletus	435		
		431	Peloponnesian War begins

CHRONOLOGICAL TABLE

Meton of Athens			
Euctemon of Athens	} 430		
Hippocrates of Chios			
Hippocrates of Cos[1]	425		
Diogenes of Apollonia	425		
		421	Peace of Nicias
		415	Athenian expedition to Sicily
Democritus of Abdera	410		
Philotans of Croton	410		
Theodorus of Cyrene	405		
		404	End of Peloponnesian War
		399	Death of Socrates
Archytas of Tarentum	385		
Philistion of Locri	385		
born 428, died	347		
Plato of Athens			
Eudoxus of Cnidus	365		
		338	Battle of Chaeronea
		336	Murder of Philip; Alexander succeeds
Callippus of Cyzicus	330		
Aristotle of Stagira			
born 384, died	322		
Heraclides of Pontus	330		
		323	Death of Alexander
Theophrastus of Eresus	320		
Strato of Lampsacus	290		

[1] The dates of the treatises in the Hippocratic Corpus cannot be fixed at all precisely. Those referred to in my text fall into three groups, (i) c. 430–380—*On Airs, Waters, Places, On Ancient Medicine, On Breaths, Epidemics* I and III, *On Fractures, On Joints, On the Nature of Man, Prognostic, On Regimen in Acute Diseases, On the Sacred Disease*. (ii) c. 400–350—*On Diseases* IV, *On Generation, On the Nature of the Child*. (iii) after 330—*On the Heart, Precepts*.

PREFACE

THE subject of this study is Greek science from its beginnings to the death of Aristotle. The term 'Greek science' and the scope of the study need a word of explanation. Science is a modern category, not an ancient one: there is no one term that is exactly equivalent to our 'science' in Greek. The terms *philosophia* (love of wisdom, philosophy), *episteme* (knowledge), *theoria* (contemplation, speculation) and *peri physeos historia* (inquiry concerning nature) are each used in particular contexts where the translation 'science' is natural and not too misleading. But although these terms may be used to refer to certain intellectual disciplines which we should think of as scientific, each of them *means* something quite different from our own term 'science'. 'Greek science' is here used, then, merely as a shorthand expression to refer to certain ideas and theories in the ancient writers, and it does not presuppose any particular view concerning the status of those ideas and theories on the part of the ancient writers themselves. Different ancient authors whom we can loosely describe as 'scientists' had, as we shall see, very different conceptions of the nature of the inquiry they were undertaking. Indeed the study of early Greek science is as much a study of the development and interaction of opinions concerning the nature of the inquiry as of the content of the theories that were put forward.

The subject-matter we shall be dealing with comprises, first, the problems, theories and methods of the various branches of science that engaged the Greeks' attention, and secondly, the ideas of the writers in question concerning the nature of the inquiry they were undertaking. But in neither case can more than a small proportion of the material be discussed: the topics I have selected come chiefly from astronomy, physics and biology, and I have included mathematics only in so far

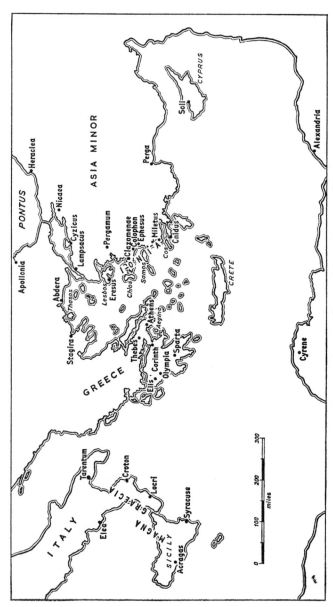

The Greek world in the fifth and fourth centuries B.C.

as it relates to these or to the development of scientific method in general.

Our sources are mainly literary and the information they provide is uneven. The evidence concerning technology and the interaction between science and technology in our period is especially poor. Our information concerning many important writers, particularly in the earlier part of the period, is derived from accounts given by later commentators which are often marred by vagueness, inconsistency or bias. On the other hand we have extensive texts from fifth- and fourth-century medical literature as well as the philosophical dialogues of Plato and most of the treatises of Aristotle. At places we have to admit the inadequacy of our sources. But some of the over-all picture is clear and we can hope to establish some of the points on which any assessment of the development of science in Greece must be based.

*   *   *

It is not possible, in a book of this scope, to acknowledge in the text all the debts owed to the work of the small, but distinguished, group of scholars who have written on early Greek science. Nor are more than a few of the most important books and articles to which I am indebted mentioned in the brief bibliography (pp. 147ff), the purpose of which is to provide a guide to further reading. I should, however, remark that, in common with most students of Greek science, I owe a special debt to the writings of four scholars in particular, namely Professors Clagett, Farrington, Neugebauer and Sambursky.

My thanks are due to Professor Moses Finley and Mr John Roberts. From the earliest stages in the writing of the book I have benefited from Professor Finley's patient and constructive criticisms and advice. Both he and Mr Roberts have read the book in draft and been responsible for numerous improvements in both style and content. It is a pleasure to record my deep appreciation of their help.

G. E. R. L.

EARLY GREEK SCIENCE
Thales to Aristotle

I

The Background and the Beginnings

It is often claimed that science began with the Greeks. What does it mean to say this? Indeed what does it mean to talk of science having an origin at all? On one view of what science is, where it is defined, as by Crowther, as 'the system of behaviour by which man acquires mastery of his environment', no human society is or ever has been without the rudiments of science. More commonly, however, science is defined more narrowly, not as a system of behaviour, but as a system of knowledge. Clagett, for example, has described it as comprising first 'the orderly and systematic comprehension, description and/or explanation of natural phenomena', and secondly, 'the tools necessary for that undertaking', including, especially, logic and mathematics.[1]

But did science, so conceived, begin at a particular point in space and time, at least so far as the Western world goes?[2] Most of those who have written on ancient science have thought so. Aristotle was the first to suggest that the inquiry into the causes of things began with Thales of Miletus. Thales and the other Milesian philosophers, Anaximander and Anaximenes, undoubtedly owed a great deal to earlier ideas and beliefs, both Greek and non-Greek, but their speculation—so it is generally argued—makes a definite break with the past, and this justifies the claim that both philosophy and science, as

[1] See J. G. Crowther, *The Social Relations of Science*, revised ed. (London, The Cresset Press, 1967), p. 1, and M. Clagett, *Greek Science in Antiquity* (London, Abelard-Schuman, 1957), p. 4.

[2] The question of the nature and extent of the similar developments that took place independently in ancient China is a complex one on which Needham's studies (*Science and Civilisation in China*, Cambridge, University Press, in progress) have thrown much light. In the present study I am concerned only with the growth of the scientific tradition in the West.

we know them, originated with them. To assess this line of interpretation we must examine just how original and distinctive the Milesians' contribution was. First, however, we must consider what there is to be said on the other side. Thales' world was far from being a primitive one, and some of the achievements of the Near Eastern civilisations with which Miletus was in contact are directly relevant to our problem.

First, technology. During the fourth and third millennian B.C. a series of extraordinarily important technological developments took place in the Nile valley and in Mesopotamia, while similar changes also occurred in the Indus valley and in China. The story has been told by such writers as Gordon Childe and Forbes.[1] The history of metallurgy stretches back to the discovery of methods of extracting metals from ores, and beyond that to the first attempts to use stone implements to work metals occurring naturally. The techniques of hammering, melting and casting were known before about 3000 B.C., and soon after alloys of copper were being produced, at first not by alloying two pure metals, but by smelting a copper ore together with an ore containing another metal or metals, tin, antimony, arsenic, lead or zinc. The processes of spinning and weaving also have their origins in prehistoric times. The skills of the ancient Egyptian textile makers can be judged from the remnants of their work that have been preserved. Some of the linen from the royal tombs at Abydos that has been dated to the First Dynasty (c. 3000 B.C.) has been found to contain 160 threads per inch in the warp and 120 in the weft. Pottery is a third invention that had far-reaching consequences for the economy of early societies. At first, pots were formed directly by hand, but the use of the potter's wheel is generally dated to about 325 B.C., and it appears that the principle of the wheel was applied only later, perhaps about 3000 B.C., to vehicles. Of even greater importance from the point of view of

[1] See the works listed in the Bibliography, Part II, Section B, pp. 149–50.

2

the growth of urban civilisation was the evolution of agriculture—the cultivation of different types of cereals and the development of techniques of irrigation and of the domestication of animals, together with the discovery of methods of preserving and preparing food. Finally, writing itself is an invention that has been dated to around the middle of the fourth millennium.

We can only guess how such technological advances were made. It is reasonable to assume that accident played a part in many discoveries. In the case of pottery, for instance, it may have been noticed that clay left by chance in a fire acquired new properties. Even so we should not underestimate the imaginative leap necessary to appreciate the potentialities of the new substance and to see how to exploit it. Compare the case of penicillin. How often, one might ask, had this mould grown on a culture dish before it was discovered by Alexander Fleming? The development of the techniques of metallurgy and of textile manufacture, especially, must each have involved a long and arduous process of learning by trial and error. Craftsmen observed the different effects produced by combining different ores in different proportions, and no doubt they often deliberately varied those proportions and tried out different techniques of smelting ores. They experimented in the general, non-technical sense of the word, their experiments being designed not to test a theory, but to improve the end-product of their work, to obtain a stronger or a tougher or a finer alloy.

The point is often made that however important technological developments were for the evolution of civilisation, they imply no science, but only guesswork and luck. But while they involve no conscious theorising, they demonstrate a highly developed ability to observe and to learn from experience. Here the evidence of anthropology can be used to supplement the findings of the pre-historians. The great French anthropologist, Claude Lévi-Strauss, in particular, has drawn attention to the complexity and minuteness of many of the classification

3

systems found in primitive societies. One example he mentions in *The Savage Mind*[1] is that of the Hanunóo of the Philippines who distinguish some 461 different zoological types, including sixty different sorts of fish and eighty-five different sorts of molluscs. Granted that the point of view from which the classes are distinguished does not correspond to that of the modern zoologist, such classifications nevertheless presuppose great skill in observation.

Technology made extraordinary advances in the fourth and third millennia. But two other features of ancient Near Eastern civilisations are more closely related to early Greek science. The first of these is medicine, the second mathematics and astronomy. Both Egyptian and Mesopotamian medicine were, it is true, dominated by magical beliefs and superstitions. Assyrian and Babylonian medical tablets show that prognosis relied heavily on divination, and in therapy the main preoccupation was to exorcise the demons that were thought to be responsible for most diseases. Medical papyri from Egypt show that there too therapy usually depended on a combination of spells and incantations and simple vegetable or mineral remedies. Yet in some respects Egyptian medicine, at least, had advanced beyond the level of folk medicine.

The famous Edwin Smith papyrus, which dates from around 1600 B.C. but incorporates material from a much earlier period, contains an account of forty-eight cases in clinical surgery, involving injuries to the head and upper part of the body. Each case report is divided into the title, the examination, the diagnosis, the treatment and explanations of difficult medical terms. The tone of the whole account is restrained. The treatments prescribed are generally simple and straightforward: 'anoint the wound with grease' is a typical example. And it is recognised that in some cases no cure is possible. This is the only such papyrus to have come down to us, but it shows that the Egyptians began at an early stage

[1] London, Weidenfeld and Nicolson, 1966, p. 4.

to attempt to record empirical data relating to particular cases in much the same way as the Hippocratic doctors were to do. Yet even in this text, which is generally quite remarkable for its freedom from magic and superstition, the writer turns at one point to supernatural aids. Case nine, which is also exceptional in containing no diagnosis and hardly any examination, ends with a description of the charm that is to be recited to ensure that the remedy is effective.

The development of mathematics and its application to astronomy were in the long run even more important than medicine. The main achievement of the Egyptians in this area was their invention of what has been described as the only intelligent calendar in human history.[1] They divided the year into 365 days, that is twelve months of thirty days each, plus five additional days, an arrangement that was far superior to the lunar, or more strictly speaking, 'luni-solar', calendars of the Babylonians and to the various, often quite chaotic, civil calendars that the different Greek city-states used.[2] These Babylonian and Greek calendars aimed to keep the months in step with the observed phases of the moon. But since the interval between one new moon and the next does not correspond to a whole number of days, the months were either twenty-nine or thirty days long. And then a much more serious complication arose from the fact that the solar year cannot be divided into an exact number of complete lunar months. Whereas the ordinary year consisted of twelve such months, with

[1] O. Neugebauer, *The Exact Sciences in Antiquity*, second edition, Providence R.I., Brown University Press, 1957, p. 81. Both the Julian and our own Gregorian calendars preserve the principle of a norm of a 365-day year, but they differ from the Egyptian in that they incorporate rules governing the introduction of an extra day in 'leap years'—thereby ensuring a closer correspondence between the solar and the calendar year.

[2] In addition to the civil calendar which regulated the religious festivals, the Athenians of the fifth century employed a 'prytany' calendar to govern the terms of office of the representatives of the ten tribes in the Council.

individual days added to some of these, in some years a whole month was 'intercalated' to keep the calendar more or less in agreement with the seasons. By the late fifth century Greek astronomers had made a tolerably accurate calculation of the number of extra months necessary in a nineteen-year cycle, and about the same time the Babylonian calendar was in fact regulated according to a predetermined schema. Yet in Greece itself, despite the advance in astronomical knowledge, the Athenian and other civil calendars remained unsystematic, the intercalation of days and months being controlled by the decision of the magistrates. Even in late antiquity Greek astronomers used the Egyptianstyle calendar for preference in their calculations.

In general, however, the Babylonians far surpassed the Egyptians in both mathematics and astronomy. First, the Babylonian number-system was based on the placevalue principle. The Egyptians, like the Romans, had separate signs for 1, 10, 100, 1000 and so on, and represented the number forty-four by four signs for ten plus four unit signs. With this we may contrast our own system, based on the place-value principle, in which the figure 11 represents not 1 plus 1, but 1 times 10, plus 1. Exactly the same principle was used by the Babylonians, although they used not ten, but sixty as their base: the unit symbol followed by the ten symbol represents the number seventy $(60 + 10)$ and so on. Some of the advantages of the place-value system become apparent when we consider how subdivisions of the unit are dealt with. In place-value notation, the operation 0.4×0.12 is handled in the same way as 4×12, whereas the equivalent operation is more complicated if we use vulgar fractions $(\frac{2}{5} \times \frac{3}{25})$. In fact the Egyptians complicated matters for themselves even further by reducing all fractions, with the exception of $\frac{2}{3}$, to fractions in which the numerator is one: they treated 0.4, for instance, not as $\frac{2}{5}$, but as $\frac{1}{3} + \frac{1}{15}$. Extensive cuneiform texts dating from the second millennium B.C. show that the Babylonians had already achieved a remarkable mastery not only in

6

the field of purely arithmetical calculations, but also in algebra, particularly in the handling of quadratic equations.

The evidence for early Babylonian astronomy is much more fragmentary. The bulk of extant astronomical cuneiform texts dates from the Seleucid period (roughly the last three centuries B.C.). But celestial omens were evidently observed and recorded from about the middle of the second millennium. One of the first such collections is that relating to the appearances and disappearances of Venus, which were recorded for several years in the reign of Ammisaduqa (c. 1600 B.C.). And by about the eighth century systematic observations of certain celestial and meteorological phenomena were being carried out for the royal court. In the second century A.D. the great Greek astronomer Ptolemy had access to fairly complete records of eclipses from the reign of Nabonassar onwards, and he used the first year of that reign (747 B.C.) as the base line for all his astronomical calculations. Such observations were originally carried out either for astrological purposes—to predict the fortunes of the kingdom or the king—or to establish the calendar—which depended on observations of the first and last visibility of the moon.

The accuracy of these early Babylonian observations should not be exaggerated: many of the phenomena they were interested in occur close to the horizon and so are very difficult to observe. Nor can the stage at which a systematic mathematical theory was applied to astronomical data be determined precisely. The chief modern authority on Babylonian astronomy, Otto Neugebauer, doubts that this was achieved before about 500 B.C. Nevertheless two positive conclusions can be drawn. First, the Babylonians had conducted extensive observations of a limited range of celestial phenomena long before Greek science began. And secondly, with the records they accumulated, they were in a position to predict certain phenomena. At no stage could they, or anyone else in antiquity, make accurate predictions of eclipses of the sun visible at any given point on the earth's surface: the most they could do here was to say when a solar eclipse was excluded or when it was possible. On the other hand they may well have been able to predict eclipses of the moon, such predictions being based not on any geometrical model of the heavenly bodies, but on purely arithmetical procedures, that is on computations from periodic tables constructed from past observations.

Yet despite the achievements of the Near Eastern peoples in the fields of medicine, mathematics and astronomy, it is still reasonable to argue that Thales was the first philosopher-scientist. We must now consider what this claim involves and how far it is justified. First it should not be supposed that what the Milesians achieved was a fully articulated system of inquiry including a definite methodology and extending over the whole of what we call natural science. Their investigations were restricted to a very narrow range of topics. They had no conception of 'scientific method' as such. It is even difficult to formulate the problems they were interested in without using such concepts as 'matter' and 'substance', although the equivalent Greek terms were not coined, let alone clearly defined, until the fourth century. Nevertheless there are two important characteristics that do distinguish the speculations of the Milesian philosophers from those of earlier thinkers whether Greek or non-Greek. First there is what may be described as the discovery of nature, and second the practice of rational criticism and debate.

By the 'discovery of nature' I mean the appreciation of the distinction between the 'natural' and the 'supernatural', that is the recognition that natural phenomena are not the products of random or arbitrary influences, but regular and governed by determinable sequences of cause and effect. Many of the ideas attributed to the Milesians are strongly reminiscent of earlier myths, but they differ from mythical accounts in that they omit any reference to supernatural forces. The first philosophers

8

were far from being atheists. Indeed Thales is reported to have held that 'all things are full of gods'.[1] But while the idea of the divine often figures in their cosmologies, the supernatural plays no part in their explanations.

A single example will illustrate this: the theory of earthquakes which is attributed to Thales. Thales apparently imagined that the earth is held up by water and that earthquakes are caused when the earth is rocked by wave-tremors in the water on which it floats. The idea that the earth floats on water is one that occurs in several Babylonian and Egyptian myths, and we have no need to go beyond Greece itself for a mythical precursor to Thales' theory, for the idea that Poseidon, the god of the sea, is responsible for earthquakes was a common Greek belief. Simple as Thales' theory of earthquakes is, it is a naturalistic explanation, making no reference to Poseidon or any other deity. First, then, to adapt an expression from Farrington, the Milesians 'leave the gods out': whereas when an earthquake or a flash of lightning is described in Homer or Hesiod, it is often, though not invariably, attributed to the anger of Zeus or Poseidon, the philosophers exclude any reference to the wills of divine personages, their loves, hates, passions and other quasi-human motives. And secondly, whereas what Homer describes is usually a particular earthquake or a particular flash of lightning, the Milesians focused their attention not on a particular example of the phenomenon, but on earthquakes or lightning-flashes in general. Their inquiries were directed towards classes of natural phenomena, and they exhibit this feature of

[1] This is reported by Aristotle (*On the Soul* 411 a 8) who is, advisedly, cautious in all his remarks concerning Thales. It is doubtful whether Thales wrote anything—certainly no written work of his was extant in Aristotle's day. In this case, then, our information takes the form of reports in Plato, Aristotle and others, attributing certain sayings or beliefs to him. For the two later Milesians, the evidence improves slightly: they definitely composed written works and it is clear that some of our sources had access to selections from these, if not to the works themselves.

science, that it investigates the universal and the essential, not the particular and the accidental.

The second distinguishing mark I referred to is the practice of debate. Admittedly we must proceed carefully here. Most of our information concerning the work of the Presocratic philosophers[1] derives from much later sources, many of which present an oversimplified picture of the continuity of early Greek speculative thought. This is especially true of the neat philosophical genealogies of the form 'A taught B, B taught C, C taught D' that frequently appear in the accounts of the doxographers—the writers who compiled collections of the opinions of the philosophers. Even Aristotle's judgements have to be treated with caution. When he suggests, for instance, that most of the Presocratics were engaged on one and the same problem, what he calls the material cause of things, we must remember that he is interpreting, not describing, the ideas of his predecessors.

Nevertheless we have reliable evidence that many of the early Greek philosophers knew and criticised one another's ideas. In many cases this can be shown by referring to the words of the philosophers themselves, where these have been preserved for us by being quoted by later writers. Thus, of the philosophers who came after Parmenides, both Empedocles and Anaxagoras adopted his principle that nothing can come to be from not-being and reformulated it in terms that echo those that Parmenides himself had used. Before that, Heraclitus mentions his predecessors and contemporaries on several occasions, notably in fragment 40[2] where he remarks sourly that 'much learning does not teach sense: for otherwise it would have taught Hesiod and Pythagoras and again Xenophanes and Hecataeus', and earlier still

[1] This term is conventionally used to refer collectively to philosophers down to and including Democritus, although strictly speaking Democritus himself was Socrates' contemporary.

[2] Original quotations from the Presocratic philosophers are referred to according to the number of the 'fragment' in the edition of Diels-Kranz (see Bibliography, Part I, Section B).

one of Xenophanes' poems is obviously making fun of the Pythagorean doctrine of the transmigration of souls —that is the belief that when a man or animal dies, its soul is reborn in another living creature. This is poem 7 where Xenophanes tells the story of how Pythagoras stopped a man beating a dog with the words: 'Stop, do not beat him. It is the soul of a friend—I recognise his voice.'

The medical writers too provide valuable supplementary evidence concerning the interchange of ideas in the late fifth century. The author of *On Ancient Medicine* objects to those medical writers who incorporate fashionable cosmological doctrines into medicine, and he refers explicitly, in Chapter 20, to the work of Empedocles. The treatise *On the Nature of Man* even describes some of the debates that were held on the question of the ultimate constituents of man. 'When the same men debate with each other in front of the same audience,' this writer says (Chapter 1), 'the same speaker never wins three times in succession,' and he mentions the philosopher Melissus by name in another connection.

Most of our first-hand evidence relates to the fifth century or later, but we may be confident that this tradition of criticism and debate goes right back to the Milesians themselves. This is clear from the nature of the rival theories they put forward on such specific topics as why —as they assumed—the earth is at rest, as well as on the major question of the origin of things in general (see below, Chapter 2). But what bearing has this tradition of criticism on the development of science? Once again we may compare the Milesians with earlier thinkers. The themes dealt with in ancient Near Eastern, or early Greek, mythology include such questions as how the world originated, how the sun travels round the earth, or how the sky is held up, but each of the myths dealing with any one of these themes is independent of the others. The Egyptians, for instance, had various beliefs about the way the sky is held up. One idea was that it is supported on posts, another that it is held up by a god, a third that it rests on walls, a fourth that it is a cow or a goddess whose arms and feet touch the earth. But a story-teller recounting any one such myth need pay no attention to other beliefs about the sky, and he would hardly have been troubled by any inconsistency between them. Nor, one may assume, did he feel that his own account was in competition with any other in the sense that it might be more or less correct, or have better or worse grounds for its support, than some other belief.

When we turn to the early Greek philosophers, there is a fundamental difference. Many of them tackle the same problems and investigate the same natural phenomena, but it is tacitly assumed that the various theories and explanations they propose *are* directly competing with one another. The urge is towards finding the best explanation, the most adequate theory, and they are, then, forced to consider the grounds for their ideas, the evidence and arguments in their favour, as well as the weak points in their opponents' theories. The Presocratic philosophers were, to be sure, still highly dogmatic: they offered their theories not as tentative or provisional accounts but as definitive solutions to the problems in question. Nevertheless they frequently show their awareness of the need to examine and assess theories in the light of the grounds adduced for them, and this principle is, one may say, the necessary precondition for progress in both philosophy and science.

But the more one argues for the originality and importance of the Milesians' contribution, the more pressing becomes the need to consider why this development took place at the time and the place that it did. This is an extremely difficult and controversial question. At one stage it would have been fashionable to refer simply to the genius of the individual philosophers, to speak of a 'Greek miracle' and to leave it at that: yet that is no explanation, but rather what we have to try to explain. On the other hand too narrow an economic explanation must also prove unsatisfactory. Certainly Miletus was, until its destruction by the Persians in 494, a rich city:

its wealth was derived partly from its industries (particularly woollen textile manufacture) and partly from trade, and it was famous as a founder of colonies. Yet while this may have been a necessary, it can hardly have been a sufficient, condition of its producing the first philosophers. The material prosperity of Miletus was, after all, no greater than that of many other cities, both Greek and non-Greek, of this period. It would be rash to claim to be able to deal with this problem adequately within the scope of this study, but some aspects of it may be noted briefly.

First we must restate precisely what has to be explained. What the Milesian philosophers achieved was, we may repeat, no fully articulated system of knowledge. Had they done so, that would indeed have rightly been considered a 'miracle'. Their achievement was rather to have rejected supernatural explanations of natural phenomena and to have instituted the practice of rational criticism and debate in that context. To understand the background to this development we must refer not only to economic factors, but also and more especially to the political conditions in Greece at the time. It is here that the contrast between the Greek world and the great Near Eastern civilisations is most marked. It is not, however, that Greece was more peaceful and stable than Lydia, Babylonia and Egypt. On the contrary, the period was one of great political upheaval throughout the Greek world, and like many other Greek cities Miletus itself suffered from bitter party strife and was ruled intermittently by tyrants. Yet whereas in the Near Eastern super-powers a change of rule usually meant no more than a change of dynasty, major developments took place in the political and social structure of the Greek cities. The seventh and sixth centuries saw the foundation and consolidation of the institutions of the city-state, the development of a new political awareness and indeed a proliferation of constitutional forms, ranging from tyranny through oligarchy to democracy. The citizens of such states as Athens or Corinth or Miletus not only often participated in the government of their country; they engaged in an active debate on the whole question of the best type of government.

This still does not help us to explain why of all the emerging Greek city-states it should have been Miletus that produced the first philosophers, and indeed in the present state of our knowledge we must admit that no definite answer to that question is forthcoming. The main features of the economic and political situation in Miletus were repeated, to a greater or lesser degree, in many other Greek cities. But while we may be no nearer explaining this phenomenon than before, we may at least see it now as part of a larger development. The freedom with which the Milesian philosophers called in question earlier ideas and criticised one another's may be compared with the spirit in which the citizens of the growing city-states debated the best form of political constitution.

A particular example may help to bring this out. Though it might seem far-fetched to compare Thales with his contemporary, the poet and law-giver Solon, the exercise reveals certain interesting points of similarity between them. First it should be noted that Thales' own activities were not confined to speculative thought. Several of the stories that are told about him relate to his engaging in business and political affairs: thus Herodotus (I, 170) reports that he advised his fellow Ionians to set up a common council and to federate. Both Thales and Solon were regularly included in the lists the Greeks drew up of their Seven Wise Men, and the Seven included a high proportion of law-givers and statesmen. Solon himself, of course, was chiefly famous for the far-reaching constitutional reforms he carried out at Athens in 594, and we are particularly lucky in having some of his own poems in which he speaks of the aims and principles that guided him. These poems indicate that he accepted personal responsibility for his proposals, and an essential item in his reform was to publish the laws and make them accessible to all Athenians. In their very different spheres of activity, the philosopher Thales and the law-giver Solon may be said to have had at least two things in common. First, both disclaimed any supernatural authority for their own ideas, and secondly, both accepted the principles of free debate and of public access to the information on which a person or an idea should be judged. The essence of the Milesians' contribution was to introduce a new critical spirit into man's attitude to the world of nature, but this should be seen as a counterpart to, and offshoot of, the contemporary development of the practice of free debate and open discussion in the context of politics and law throughout the Greek world.

2

The Theories of the Milesians

THE general character and significance of Milesian
speculation have been discussed in my opening chapter.
It is now time to consider in greater detail some of the
specific theories and explanations they put forward. Our
material falls into two groups, first, theories dealing with
particular phenomena or problems, such as the nature of
thunder and lightning, or what the stars are made of, or
why the earth is at rest, and secondly, doctrines of a
general cosmological import. The fact that so many
theories of the first kind have been reported is partly due
to the interests of the doxographical sources on whom we
depend. But it is clear that the Milesians paid a good
deal of attention to rare or striking natural phenomena.
If we ask why this was so, part of the answer may lie in
their desire to provide naturalistic explanations for
phenomena that were usually considered to be controlled by the gods. Zeus was responsible for the thunder-
bolt and Poseidon for earthquakes, and Atlas held the
earth up on his shoulders. The bid to give naturalistic
explanations was going to be judged especially by its
success or failure to account for events that were popu-
larly believed to be produced by supernatural agencies.
Thus Thales, as we have already noted, explained earthquakes as occurring when the earth is rocked on the
water on which it floats. Similarly Anaximander suggested that thunderbolts are caused by wind and that
lightning is produced when clouds are split in two. Naïve
as these explanations are, their significance lies not so
much in what they include as in what they omit, that is
the arbitrary wills and quasi-human motivation of the
anthropomorphic gods.

Moreover in some cases the Milesians went far beyond
merely countering popular beliefs in the supernatural.

16

Two examples from Anaximander are particularly interesting. First he attempted an account of the heavenly
bodies, picturing them as rings of fire. The rings themselves cannot be seen since they are surrounded by mist,
but they have openings through which the heavenly
bodies appear: what we see as a star is like a puncture in
a vast celestial bicycle wheel. He postulated three such
rings for the sun, the moon and the stars: the diameters
of the rings are twenty-seven, eighteen and nine times
the diameter of the earth, the earth iself being represented as a flat-topped cylinder, three times as broad as
it is deep, at rest at the centre of the rings, and he suggested that eclipses of the sun and moon occur when the
apertures through which they are seen become blocked.
Many difficulties remain: how the circle or sphere of the
fixed stars is conceived is far from clear; nothing is said
about the planets as such; most strangely, the fixed stars
are held to be below both the sun and the moon. Nevertheless the importance of this theory is that it is the first
attempt at what we may term a mechanical model of the
heavenly bodies in Greek astronomy.

My second example is the theory Anaximander proposed concerning the origin of animals in general and of
man in particular. This too was a topic on which several
myths were current in ancient Greece. There was the
story of Deucalion and Pyrrha, for example, who, when
the flood had wiped out the human race, created more
men and women by throwing stones over their shoulders.
In other stories mankind was represented as related to
and derived from the gods. As we should expect, Anaximander's approach to this subject is quite different.
According to a report in the third-century A.D. doxo-
grapher Hippolytus, he held that living creatures are
first generated in the 'wet' when this is acted on by the
sun. No doubt, like most Greeks, he believed that
animals may be spontaneously generated in certain substances under certain conditions, and this idea provided
the basis for an account of the origin of animals as a
whole. But he also suggested that man was originally

17

born in a different species of animal, that is, apparently, some sort of fish. Another of our sources, Plutarch (*Table Talk* VIII, 8, 4, 730e), refers in this context to the *galeoi* or dog-fish, one species of which, the so-called 'smooth shark', is remarkable in that the young are attached by a navel-string to a placenta-like formation in the womb of the female parent. Many scholars have considered it unlikely that Anaximander himself knew of that particular species, but if Plutarch's report has any substance at all, it can hardly be doubted that Anaximander had some knowledge of viviparous sea-animals of some sort.

But the question of what empirical basis, if any, Anaximander's theory may have had is less important than the reasoning that led him to propose it in the first place. The original stimulus seems to have been the observation that on birth the human infant takes a long time to become self-sufficient. Anaximander seems to have appreciated that this created a serious difficulty for anyone who supposed that the human race originated with the sudden appearance, on earth, of the young of the human species. He preferred to argue that humans must originally have been born from a different species of animal which was able to nurture them long enough for them to become self-sufficient. Neither Anaximander nor any other Greek theorist developed a systematic theory of the evolution of natural species as a whole. But as this example shows, the Greek philosophers began at an early stage to reflect on the problems posed by the origin of the human race and by man's development from nature to culture.

The three main theories attributed to the Milesians are their more general cosmological doctrines which Aristotle interpreted as relating to the 'material cause' of things. Thales is supposed to have held that this is water, Anaximander the 'Boundless', and Anaximenes air. It seems more likely, however, that they were not concerned with precisely the same problem, but rather with three slightly different versions of it. What type of question, we might ask, could Thales have put to him-

18

self? Certainly not the question suggested by Aristotle, at least not in the terms he uses when he comments in the *Metaphysics* (983 b 6 *ff*) that

> most of the first philosophers . . . thought the principles which were of the nature of matter were the only principles of all things. That of which all things are made, from which they first come to be, and into which they are ultimately resolved (the substance persisting, but changing in its attributes), this they say is the element and this the principle of things.[1]

The terms translated as 'matter', 'substance', 'attribute' and 'element' were all introduced for the first time into philosophy in the fourth century, and it is inconceivable that any of the Milesians used them.

On the other hand there was clearly nothing to prevent Thales asking, for example, what the origin or beginning of things was. The poet Hesiod, after all, had already proclaimed in the *Theogony* (116) that 'first of all chaos [that is the yawning gap] came to be', and he had gone on to describe how the gods and other personified figures came to be, relating them all together in one vast family tree. So Thales may well have asked what the origin of things was in the sense of what came first— although the answer he gave to this question differed fundamentally from Hesiod's in that it referred not to a mythical 'yawning gap', but to an ordinary substance, water.

But if this is a problem that Thales might have considered, indeed almost certainly did consider, it is an open question whether he *also* asked whether or in what sense the substances in the world around us still consist of water. Did he believe that the chair he sat on, and the bread he ate, were made of water? Not much later than Thales, Anaximenes certainly held some such belief,

[1] Based on the Oxford translation, *The Works of Aristotle translated into English*, edited by W. D. Ross (Oxford, Clarendon Press), *Metaphysics*, W. D. Ross (Vol. VIII, 2nd ed., 1928).

although his originative substance was air, not water. However, as we shall see, Anaximenes offered a definite account of the changes that air undergoes to appear, for example, as earth or stone. Our sources are silent about how Thales would have explained this. We cannot tell for certain whether this is simply because of the gaps in the information available to us, or because Thales never considered the problem. But when we piece together the admittedly fragmentary evidence for the development of Milesian speculation, and particularly the evidence relating to Anaximander, the latter explanation is more likely. Despite the testimony of Aristotle, it is probable that while Thales both asked and answered the question of what came first, the question of how or in what sense the primary substance persists in the objects we see around us only arose as the result of further inquiry.

Anaximander suggested that the first thing was not any specific substance, but something indefinite, which he named the Boundless. If we ask why he chose this, rather than a familiar substance such as water, a passage in Aristotle (*Physics* 204 b 24 *ff*) helps to suggest a plausible answer. Anaximander may have appreciated one difficulty that theories such as Thales' must run into, namely this: how, if the primary substance is water, for example, can its opposite, fire, ever have come to be, for each destroys the other? If this was indeed Anaximander's reasoning, it provides a good illustration of what I called earlier the practice of rational criticism, and elsewhere too his theories seem to spring from a realisation of possible objections to those of his predecessor. A second striking instance concerns his theory of what holds the earth up. Where Thales had suggested that the earth floats on water, Anaximander held that it 'hangs freely', 'remaining where it is because of its equal distance from everything', as Hippolytus puts it (*Refutation of All Heresies* I, 6, 3). Here too the impulse to put forward this startlingly sophisticated doctrine may well have been that he appreciated that Thales' view, and

20

views like it, run into an obvious difficulty: if water holds the earth up, what holds the water up?

But our evidence for Anaximander also contains an account of how the Boundless developed, and this is important for the light it throws on the relation between him and the other two Milesians. According to one source (the *Stromateis* or *Miscellanies* attributed to Plutarch, Chapter 2) he suggested that 'at the birth of this world a germ [or seed] of hot and cold is separated off from the eternal [that is, the Boundless] and from this a sphere of flame grew round the air surrounding the earth, like bark round a tree'. Thales, I suggested, may well not have considered the question of what happened to his primary substance, water. But we can be fairly certain that Anaximander put forward what we should call a cosmogonical theory. Roughly speaking, his main suggestion was that the cosmos grows like a living thing from a seed. What is particularly interesting about this biological model is that it might well allow Anaximander to side-step the question of whether the substances we see around us are the same as, or different from, the original substance from which they came. Take the growth of a plant. The apparently homogeneous seed gives rise to a great many different things, leaves, fruit, roots, bark and so on. Admittedly an Aristotle would ask the same question here as of the world as a whole: are these things new substances, or qualitative modifications of the original substance? But if Anaximander believed that different things arise from the Boundless just as *naturally* as all the parts of a tree from the seed from which it grows, it may be that he did not consider that question. While he seems to have gone further than Thales in offering an account of the development of the world, he too, like Thales, may have had no definite view on the question of whether this piece of wood or that piece of bread is the same in substance as the Boundless.

If this interpretation is sound, it was not until Anaximenes, the third of the Milesian philosophers, that this last question came to the fore. He too lacked the technical vocabulary to refer to the 'qualitative' modifications of the underlying substance or 'substratum'. But that did not prevent him from putting forward a clear account of the changes that affect the primary substance. In his view the primary substance was air, and at first sight this looks a retrograde step, a return to a tangible first substance like Thales' water, after Anaximander's more imaginative postulate. But the important point is that Anaximenes combined a theory of what things came from with a definite suggestion about how they came from it, namely by a process of rarefaction and condensation. The precipitation of rain illustrates how 'air' condenses to form water, and water in turn condenses to form the solid, ice; and conversely 'air', for example, is formed by rarefaction from water when it evaporates or is boiled. These simple and obvious changes provide the basis of Anaximenes' generalisation that all things come from one primary substance by a single two-way process of condensation and rarefaction. Unlike Anaximander's brilliant but arbitrary conception of the world growing from the undifferentiated Boundless, Anaximenes' theory referred to processes that can still be observed at work in natural phenomena.

The history of Milesian views about the primary substance is chiefly remarkable for the way in which the awareness of the problems grew from one philosopher to the next. Anaximander's suggestion that the primary substance is undifferentiated seems to counter an obvious objection to Thales' postulate of water—that is, how can its opposite, fire, have come to be? Anaximenes' theory of condensation and rarefaction gives a clearer account of the changes that affect the primary substance than Anaximander's idea that a seed separates off from the Boundless. As is usual in the history of science, their actual theories strike a later age as childish—they already appeared so to Aristotle. But the measure of their achievement is the advance they made in grasping the problems. They rejected supernatural causation and appreciated that naturalistic explanations can and should be given of a wide range of phenomena: and they took the first tentative steps towards an understanding of the problem of change.

3

The Pythagoreans

THE speculative thinkers of the sixth and fifth centuries are collectively known as the Presocratic philosophers, but the fact that we apply the same term 'philosopher' to all these men should not be allowed to obscure the important differences between them, for they had very different aims and interests and indeed very different social roles. There are several striking contrasts between the Milesians and the thinkers we must next consider, the so-called Pythagoreans, and the Pythagoreans themselves were far from being a homogeneous group.

Very little is known for certain about Pythagoras himself. We gather that he was born in Samos some time before the middle of the sixth century and that he later moved to Croton in Magna Graecia[1] to escape the tyranny of Polycrates in Samos. The followers of Pythagoras tended out of piety to ascribe their own ideas to the founder himself, and when our late sources do the same, they must be treated with caution. Nevertheless we have it on good authority that Pythagoras taught a way of life—for that is what Plato tells us in the *Republic* (600ab). The early Pythagoreans were not only, and not even primarily, interested in the inquiry concerning nature. They were a group held together by religious beliefs and practices. Thus they believed in the immortality and transmigration of souls, and they practised certain ritual abstentions, for example from certain types of food. Moreover they acted together as a political force in several cities in Magna Graecia in the late sixth century.

Here, then, was one way in which the Pythagoreans

[1] This is the term applied to the area of what is now southern Italy that was colonised and controlled by the Greeks from the late eighth century.

differed from the Milesians. And another is in the type of cosmological theory that some of them put forward. Where Aristotle represents the Milesians as speculating about the 'material cause' of things, he has this to say about the chief doctrines of the Pythagoreans (as his opening words show, he is referring to Pythagoreans of the fifth century rather than to Pythagoras' own contemporaries):

> Contemporaneously with these philosophers [Anaxagoras, Empedocles and the atomists] and before them, the so-called Pythagoreans, who were the first to engage in mathematics, advanced this study, and being trained in it they thought that its principles were the principles of all things. But of these principles numbers are by nature the first, and in numbers they seemed to see many resemblances to the things that are and come to be—more than in fire and earth and water . . . ; and again they saw that the modifications and the ratios of the musical scales were expressible in numbers. Therefore, since all other things seemed in their whole nature to be modelled on numbers, and numbers seemed to be the first things in the whole of nature, they supposed the elements of numbers to be the elements of all things, and the whole heaven to be a musical scale and a number (*Metaphysics* 985 b 23 *ff*).[1]

According to Aristotle, these Pythagoreans found the principles of all things in numbers. Where the Milesians had chosen material substances as the primary things—for even Anaximander's Boundless is material, just as much as Thales' water or Anaximenes' air—the Pythagoreans may be said to have focused attention on the formal aspect of phenomena. Whether or not they were the first to recognise the numerical ratios of musical harmonies, this certainly provided one of their chief examples to illustrate the role of number. The intervals of an octave, fifth and fourth could all be expressed in

[1] Based on the Oxford translation, *The Works of Aristotle translated into English*, edited by W. D. Ross (Oxford, Clarendon Press), *Metaphysics*, W. D. Ross (Vol. VIII, 2nd ed., 1928).

terms of simple numerical ratios, 1:2, 2:3 and 3:4. Here was a startling instance of phenomena that had no obvious connection with numbers exhibiting a structure that could be expressed mathematically, and it seemed to the Pythagoreans that if this applied to musical intervals, it might well be true of other things too, if only their mathematical relations could be discovered.

The importance of this search for numbers in things is clear. The Pythagoreans were thus the first theorists to have attempted deliberately to give the knowledge of nature a quantitative, mathematical foundation. This places them at the head of what was to be a development of the greatest importance for science. But to put their achievement into perspective we must add two things. The first is that the Pythagoreans held not merely that the formal structure of phenomena is expressible in numbers, but also that things consist of numbers: many of them assumed that things are made of numbers, the numbers themselves being conceived as concrete material objects.

Secondly, many of the resemblances that the Pythagoreans claimed to find between things and numbers were quite fantastic and arbitrary. Thus we are told that they equated justice with the number four (the first square number) and marriage with the number five (this represents the union of male—identified with the number three—and female—two). Opportunity, apparently, was identified with the number seven, and the special significance attached to this number evoked some sharp criticisms from Aristotle:

> Why need these numbers be causes? There are seven vowels, the scale consists of seven strings, the Pleiades are seven, at seven animals lose their teeth (at least some do, though some do not), and the champions who fought against Thebes were seven. Is it then because the number is the kind of number it is, that the champions were seven or the Pleiad consists of seven stars? Surely the champions were seven because there were seven gates or for some other

reason, and the Pleiad *we* count as seven, as we count the Bear as twelve, while other peoples count more stars in both. . . . These people are like the old-fashioned Homeric scholars, who see small resemblances but neglect great ones (*Metaphysics* 1093 a 13 *ff*).[1]

Obviously while the search for numerical ratios proved fruitful in such fields as the analysis of musical harmonies, and mathematics itself, it also and more often led to mumbo-jumbo and crude number-mysticism.

One of the examples that Aristotle gives of the arbitrary manipulation of numbers by the Pythagoreans is from astronomy, and their speculations in this field deserve more detailed consideration. Here too they were much influenced by religious and ethical motives. They believed that the whole heaven is 'a musical scale and a number', and according to the famous doctrine of the harmony of the spheres,[2] the movements of the heavenly bodies give rise to concordant, though inaudible, sounds: the reason why we do not hear them, according to one report, is that we have been used to them since birth. Moreover the soul was also conceived as a *harmonia* or attunement, and its welfare depends on its being well-tuned and orderly, *kosmios*, like the world-order or cosmos itself.

Yet these doctrines certainly did not prevent, and probably even encouraged, Pythagorean speculation about the relations between the heavenly bodies. Several different theories are attributed either to the Pythagoreans as a whole, or to different groups or individuals among them. Thus in one doctrine which is generally

[1] From the Oxford translation, *The Works of Aristotle translated into English*, edited by W. D. Ross (Oxford, Clarendon Press), *Metaphysics*, W. D. Ross (Vol. VIII, 2nd ed., 1928).

[2] The Pythagoreans and many later Greek astronomers imagined the visible heavenly bodies as situated on, and carried round by the movement of, concentric spheres that are themselves invisible. There is one sphere to each of the planets, the sun and the moon, and a single sphere for the stars (often called, in Greek astronomy, the 'fixed' stars to contrast them with the 'wandering' stars or planets).

27

taken to represent an early Pythagorean tradition, the earth is at the centre of the universe and it contains a fiery core, 'Hestia', the central 'hearth'. But a second theory is also reported and attributed by some of our post-Aristotelian sources to Philolaus of Croton, a late fifth-century Pythagorean, in particular. In this, Hestia, the central fire, is not within the earth, but is a separate body, and the earth itself is imagined as circling round it like the other heavenly bodies, the planets, sun and moon. This system is, then, neither geocentric, nor yet heliocentric. The centre is an invisible body of fire, and the doctrine is further complicated by the introduction of a second invisible body, the 'counter-earth', which circles the central fire underneath the earth. Reading from the centre outwards, then, we have the central fire, next the counter-earth, then the earth itself, and outside the earth, the moon, the sun and the planets.

The main evidence for this theory comes from two passages in Aristotle which severely criticise the grounds on which it was put forward. In *On the Heavens* (293 a 17 ff) he says:

> Concerning the position [of the earth] there is some divergence of opinion. Most of those who hold that the whole universe is finite say that it lies at the centre, but this is contradicted by the Italian school called Pythagoreans. These affirm that the centre is occupied by fire, and that the earth is one of the stars, and creates night and day as it travels in a circle about the centre. In addition they invent another earth, lying opposite our own, which they call by the name of 'counter-earth', not seeking accounts and explanations in conformity with the appearances, but trying by violence to bring the appearances into line with accounts and opinions of their own.[1]

Another highly critical comment on the Pythagorean theory occurs in a passage in the *Metaphysics* (986 a 3 ff):

> All the properties of numbers and scales which they

[1] From the Loeb translation by W. K. C. Guthrie (Cambridge, Mass., Harvard University Press; London, Heinemann, 1939).

28

could show to agree with the attributes and parts and the whole arrangement of the heavens, they collected and fitted into their scheme; and if there was a gap anywhere, they readily made additions so as to make their whole theory coherent. For example, as the number ten is thought to be perfect and to comprise the whole nature of numbers, they say that the bodies which move through the heavens are ten, but as the visible bodies are only nine [that is the sphere of the fixed stars, counted as one, plus the five planets, sun, moon and earth], to meet this they invent a tenth—the 'counter-earth'.[1]

Aristotle dismissed the doctrine of the counter-earth as a piece of fanciful number-mysticism, but another passage in *On the Heavens* (293 b 23 *ff*) suggests that this is not quite the whole story, for there he indicates that the theory was brought to bear on a genuine difficulty, namely why eclipses of the moon are more frequent than those of the sun. Although, if we take the earth as a whole, solar eclipses are more common, only a small proportion of these can be observed from any particular place. On an average, eclipses of the moon visible at any one place are about twice as frequent as eclipses of the sun, and the Pythagoreans apparently tried to account for this by suggesting that not only the earth, but also the counter-earth, intervenes between the moon and its source of light. However, the details of this theory remain, like much else in their astronomy, both vague and obscure, and they evidently made no attempt to give a precise mathematical account of the relations between the heavenly bodies.

Undoubtedly the most interesting feature of the system we have outlined is that it removed the earth from the centre of the universe. Moreover it did so, in large part, for symbolic reasons. According to yet another passage in Aristotle (*On the Heavens* 293 a 30 *ff*),

[1] From the Oxford translation, *The Works of Aristotle translated into English*, edited by W. D. Ross (Oxford, Clarendon Press), *Metaphysics*, W. D. Ross (Vol. VIII, 2nd ed., 1928).

the earth was not considered *noble enough* to occupy the most important position in the universe. Whereas religious considerations weighed with some Greek theorists *against* shifting the earth from the centre of the universe, they could and sometimes did weigh *for* that very conclusion. Whatever later astronomers felt, some of the Pythagoreans clearly had no compunction in removing the earth from the centre and making it subject to movement like a planet.

Two further aspects of the work of the Pythagoreans throw light on the methods of early Greek science, (i) the evidence concerning their empirical investigations in acoustics, including the use of simple experiments, and (ii) the development of deductive methods in mathematics. In both cases the information available from our sources leaves much to be desired, and in both cases it relates mainly to thinkers active in the late fifth or early fourth century.

Pythagoras' 'discovery' of the ratios of the musical harmonies was the subject of many legends in antiquity, several of which purport to describe how he arrived at his conclusion by carrying out observations or simple experiments, as for example by noticing the relation between the weights of hammers which made different notes when struck, or by filling jars with varying amounts of water and noting a relation between the quantity of the water and the sound the jar made when struck. Most of these stories must be rejected for the simple reason that the operations they describe do not, in fact, yield the results that are reported. But not all the accounts are fanciful. The stories that refer to his measuring the lengths of string that gave different notes, or to his making similar measurements of the columns of air in pipes, are more plausible and may well reflect the type of empirical investigations that were undertaken by Pythagoreans in the late fifth and early fourth centuries. Archytas of Tarentum, in particular, collected a variety of evidence in an attempt to establish his theory of the relation between the pitch of a note and its 'speed': in fragment 1 one of his simpler examples refers to the different notes made by different lengths of pipe in a flute. And when Plato too refers to early experiments in acoustics, his testimony is all the more convincing as he himself disapproved so strongly of this method of dealing with the problems. In the *Republic* (531a–c) he makes Socrates speak contemptuously of those who 'measure the harmonies and sounds they hear against one another', who 'torture and rack the strings on the pegs' and 'look for numbers in these heard harmonies'. All this is far from showing that the Pythagoreans recognised the value of the experimental method in general. But it does suggest that some of them carried out certain simple experiments in one field, acoustics, at least. There is, however, little need to emphasise that the motive for which they carried out those experiments was a special one, namely to support the doctrine that 'all things are numbers' by revealing the numerical relations underlying the phenomena.

The history of mathematics in the period before Plato is obscure. The reliable first-hand evidence is scanty, and widely divergent views have been taken on the extent to which our first major mathematical text, the *Elements* of Euclid (composed about 300 B.C.), was based on earlier work. Down to the mid fifth century the Pythagoreans seem to have been chiefly interested in certain aspects of the theory of numbers. The classification of numbers as odd and even probably dates from that period, and so too does the association of certain numbers with geometrical figures of different kinds; thus 4 and 9 are 'square' numbers, 6 and 12 'oblong' ones (where the sides, i.e. factors, differ by 1) and so on (see diagram 1). No doubt the early fifth-century mathematicians were also familiar with certain simple geometrical theorems, including the theorem called after Pythagoras himself, namely that the square on the hypotenuse of a right-angled triangle is equal to the sum of the squares on the other two sides: indeed the truth of that theorem had long been known to the Babylonians, for series of

'Pythagorean' numbers, such as 3, 4, 5, are recorded in cuneiform texts from the second millennium. In such cases the distinctive Greek contribution was not the discovery of the theorem, so much as its proof. But whether or to what extent such proofs were attempted before the middle of the fifth century is not at all clear: on the most

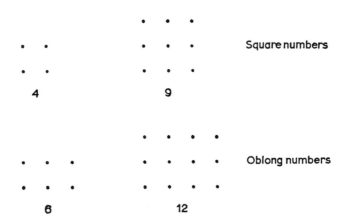

Diagram 1   Pythagorean 'square' and 'oblong' numbers.

likely interpretation of the evidence, the development of methods of mathematical demonstration is a product of the late fifth or early fourth century, and it is one that should undoubtedly be connected with other mathematicians besides those who may be considered Pythagoreans. While the detailed history of this development cannot be undertaken here, two examples may be mentioned briefly in order to illustrate the problems and methods of Greek mathematics before Plato.

My first example illustrates both the uncertainties of the evidence relating to Greek mathematics and what I have referred to as one of their distinctive contributions. This concerns the irrationality of $\sqrt{2}$ —the fact that its value cannot be expressed as a proportion between two integers—or, to put it, as the Greeks generally did, in geometrical terms, the fact that the diagonal of a square is not commensurable with its side. Approximations to the value of $\sqrt{2}$ are already found in Babylonian mathematical texts. What the Greeks did, at some stage in the late fifth or early fourth century, was to demonstrate its irrationality. The traditional proof, which is alluded to by Aristotle (*Prior Analytics* 41 a 23 *ff*), proceeds by first assuming that the diagonal of a square is commensurable with its side, and then showing that this assumption leads to the impossible consequence that the same number is both odd and even (see the additional note at the end of this chapter). Unfortunately we have no means of determining when this proof was discovered, nor even when the fact of the irrationality of $\sqrt{2}$ was known to the Greeks. Most of the stories that we find in our sources on this topic are late fabrications, for example the legend that has it that an anonymous Pythagorean, usually taken to be Hippasus, divulged this secret and suffered death by drowning as divine retribution for doing so. We do not even know whether this discovery was made as a result of exploring the applications of Pythagoras' theorem, or whether—as has recently been thought more likely—it was prompted by philosophical problems relating to the idea of infinite divisibility. The only safe conclusion that our evidence allows us to draw is that the irrationality of $\sqrt{2}$ was known before the time of Plato. In the *Theaetetus* (147d) the mathematician Theodorus of Cyrene is described as 'showing that the sides [i.e the roots] of the squares representing three square feet and five square feet are not commensurable in length with the line representing one foot', and taking all the cases up to the side of seventeen square feet in turn. While it is interesting that the problem of irrationals is still not treated here as a general problem, and is handled geometrically rather than arithmetically, the text clearly implies a familiarity with *some* demonstration of the incommensurability of the side and the diagonal of the square representing two square feet, since this fact is assumed as requiring no proof.

In my second example the evidence is more certain and the role of a Pythagorean mathematician more definite. One of the problems that exercised Greek mathematicians from the middle of the fifth century was that of the duplication of the cube: given a particular cube, how does one construct a cube of twice its volume? According to our sources Hippocrates of Chios, who is not to be confused with his contemporary and namesake, the great physician of Cos, recognised that this problem is equivalent to that of finding two mean proportionals (a, b) between two given lengths (x, y) such that $x:a = a:b = b:y$. This will provide the solution, since in the particular case where $y = 2x$, the cube on a will be double the cube on x. But the first to solve the problem of finding two mean proportionals was the Pythagorean Archytas, whose work in acoustics has already been mentioned. His solution, which has come down to us in a commentary on Archimedes, is a geometrical one and remarkable for its ingenuity. Some idea of this may be gathered from the opening words of Heath's account.[1] Heath describes it as a

> bold construction in three dimensions, determining a certain point as the intersection of three surfaces of revolution, (i) a right cone, (ii) a cylinder, (iii) a *tore* or anchorring with inner diameter *nil*. The intersection of the two latter surfaces gives (says Archytas) a certain curve . . . and the point required is found as the point in which the cone meets this curve,

—whereupon Archytas demonstrates how the point so determined enables the two mean proportionals to be found. This example indicates the progress that had already been made in geometry in the early fourth century: Archytas' brilliant three-dimensional kinematic

[1] *A History of Greek Mathematics*, Vol. I, Oxford, Clarendon Press, 1921, pp. 246 *ff*.

construction provides a foretaste of the methods that were to lead to one of the most remarkable achievements of early Greek science, the astronomical model of Eudoxus.

Additional Note

The 'traditional' proof that the diagonal of the square is incommensurable with the side is given in an appendix to Euclid, book X, and may be paraphrased as follows:

Let AC be the diagonal of the square, AB its side.

Suppose AC is commensurable with AB, and let $a:b$ be their ratio expressed in the lowest terms. Since $AC > AB$, $a > 1$.

Then $AC:AB = a:b$

So $AC^2:AB^2 = a^2:b^2$

But (by Pythagoras' theorem) $AC^2 = 2AB^2$

Therefore $a^2 = 2b^2$

So a^2, and therefore a, is even, and since $a:b$ is in its lowest terms, b is odd.

Since a is even, let $a = 2c$

So $4c^2 = 2b^2$

So $2c^2 = b^2$

From which it follows that b is even.

Since the assumption that AC is commensurable with AB leads to the impossible consequence that the same number (b) is both odd and even, the assumption must be false.

4

The Problem of Change

THE beginnings of an awareness of the problem of
change can be traced back to Milesian speculation about
the primary substance, outlined in Chapter 2. In the
early fifth century this became the chief problem in the
inquiry concerning nature. The Milesians had taken it
for granted that change occurs and that the world of
sense-experience is no illusion. But soon afterwards
philosophers began to question the basis of our knowledge of the external world. Can we trust the senses, or
should we rely on reason alone? Changes certainly seem
to occur, but do the appearances correspond to the
underlying reality or are they a misleading guide? Once
these questions had been raised, any investigator who
wished to tackle the problem of the ultimate constituents of matter had first to consider certain preliminary,
but fundamental, philosophical issues. He could no
longer take common sense for granted, but had to give
an account both of the foundations of knowledge (the
problem of epistemology) and of the nature of change
and coming-to-be.

The first philosophers to raise these questions were
Heraclitus and Parmenides, the one an Ionian from
Ephesus, the other a native of Elea, a Greek colony on
the west coast of Italy, south of Naples. We do not know
for sure whether either of these brilliant and highly
original thinkers influenced the other, although it has
been thought likely that Parmenides was acquainted
with the work of Heraclitus. What is certain is that some
time in the early part of the fifth century they both
raised the problem of change in an acute form and presented diametrically opposed solutions to it, for, whereas
Heraclitus claimed that everything is subject to change,
Parmenides denied that change can occur at all.

36

THE PROBLEM OF CHANGE

The interpretation of Heraclitus' position is disputed. Most ancient critics, beginning with Plato and Aristotle, assumed that he believed that every single thing in the world is constantly in change, but many modern commentators have argued that the thesis he was proposing was a much weaker one, namely that the world as a whole is in continuous change—that each individual thing is subject to change at some time or another. The evidence does not allow us to settle this question finally. The famous dictum *panta rhei*, 'everything flows', cannot definitely be ascribed to Heraclitus, and even if it could, this would not solve the problem, since the point at issue is whether this dictum is to be taken literally. It is, however, agreed that Heraclitus wished to emphasise the changes and interactions taking place in the world at large. Change is confined within certain limits or 'measures' which ensure a balance between the things that interact. But it is evidently an important part of his message that a state of apparent rest or equilibrium may conceal an underlying tension or interaction between opposites, and this is illustrated, in his fragments, by such examples as the strung bow or lyre, which while they seem at rest are in fact in a state of tension.

On the problem of knowledge, Heraclitus did not reject the evidence of the senses entirely, but stressed that it must be used with caution. One fragment (107) warns that 'eyes and ears are bad witnesses for men if they have souls that cannot understand their language'. But Parmenides based his philosophy on a far more radical view of the foundations of knowledge. In fragment 7 he says: 'do not let habit, born of experience, force you to let wander your heedless eye or echoing ear or tongue along this road, but judge by reason. . . .' Here he goes far beyond Heraclitus or any earlier philosopher in insisting that reason alone is to be trusted and that the evidence of the senses is utterly unreliable and misleading.

The first part of Parmenides' philosophical poem[1] is

[1] Although the Milesians Anaximander and Anaximenes, the Ephesian Heraclitus and most of the later Presocratics chose prose

devoted to what he calls the Way of Truth. In this he explores what follows from the single statement 'It is'. The starting-point of his argument, as he expresses it in fragment 2, is the statement 'It is and it cannot not be'. The subject of this sentence is not specified, and, depending on how we take it, the statement is *prima facie*, at least, open to several different constructions. Parmenides is evidently asserting the existence of something, but this may be either (*i*) being or existence itself, or (*ii*) 'what is' in the sense of all that is, i.e. the totality of existing things, or (*iii*) 'what is' in the sense of anything that is, i.e. any particular object that exists, or (*iv*)—if we take fragment 2 with other statements of Parmenides—'what can be spoken or thought about'. But while the starting-point of his argument is left vague, the conclusions he reaches by the end of the Way of Truth are perfectly explicit. He goes from a position in which he appears to deny merely that anything can come to be from the totally non-existent, to denying that anything can come to be in any sense at all. Although the second part of his poem, the Way of Seeming, contains a cosmogony, this implies no modification in the positions he advanced in the Way of Truth. On the contrary, the Way of Seeming is described as 'deceitful' (fragment 8, v. 52), no doubt because it is based on what he had earlier shown to be a radically mistaken view of being and not-being. The Way of Truth declares that coming-to-be, passing-away and change of any sort are all alike impossible.

After this devastating attack on the notion of change, any theorist who wished to put forward a physical or cosmological doctrine had first to come to terms with

as their medium—Anaximander being not only the first philosophical writer, but perhaps also the first Greek writer of any kind, to do so—three important Presocratic thinkers—Xenophanes, Parmenides and Empedocles—wrote in verse. Both Parmenides (fragment 1) and Empedocles (fragment 3) claimed divine inspiration for their philosophies, and in both cases it would be rash to discount this as merely a matter of convention.

Parmenides' arguments and with the theory of know-ledge on which they were based. The history of late fifth-century speculative thought is largely one of the controversy between those who supported Parmenides and those who resisted his conclusions. Parmenides' own followers, the so-called Eleatics, Zeno of Elea and Melissus of Samos, accepted his position whole-heartedly and developed further arguments to refute the ideas of plurality and change. But in the opposite camp the most important of the 'physicists'—that is the philosophers of nature, *physis*—also took their starting-point from Parmenides. Thus both Empedocles of Acragas and Anaxagoras of Clazomenae endorsed Parmenides' dictum that nothing can come to be from not-being, and, as we shall see, the atoms of Leucippus and Democritus have several characteristics in common with the one unchanging being of Parmenides' Way of Truth. How to counter Parmenides' denial of change was, indeed, the chief preoccupation of each of these later Presocratic systems.

Where Parmenides insisted on relying on reason alone, Empedocles reinstates the senses. He concedes that they are feeble instruments, but so too is the mind, and we should use every means at our disposal, includ-ing sight, hearing and the other senses, to grasp each thing (fragments 2 and 3). Fragments 12–14 echo Parmenides' statement that nothing can come to be from not-being, but Empedocles restores the notion of change by denying the uniqueness of what exists. Earth, water, air and fire all exist and have always existed, and they produce change by mixing with, and separating from, one another under the influence of the two oppos-ing forces that Empedocles calls Love and Strife. Nothing comes from the non-existent. But change can and does occur, this being interpreted as the mixture and separation of already existing substances.

From the point of view of the history of scientific theories, two features of Empedocles' system are par-ticularly important: his conception of physical elements, and his use of the idea of proportion. The term 'element' is ambiguous, being used (*i*) of 'original' substances—substances that have existed as long as anything has existed—and (*ii*) of 'simple' substances—the substances into which compound things can be analysed, but which themselves cannot be further reduced. Traces of both ideas can be found long before Empedocles. Thales' water, Anaximander's Boundless and even perhaps Hesiod's 'yawning gap' are 'elemental' in the first sense. And the idea that certain complex things are made of other simpler ones also occurs very early in Greek thought in certain contexts. Thus the belief that human beings are made of 'earth' and 'water' is a popular one which is implicit in, for example, the myth of Pandora in Hesiod (*Works* 59 *ff*): Hephaestus makes Pandora, the first woman, by taking earth and mixing it with water and moulding it into shape. Xenophanes of Colophon repeats the idea that human beings are made of earth and water in a non-mythical account, and in the cosmology put forward by Parmenides in the Way of Seeming everything is derived from a pair of principles, Light and Night.

But Empedocles expresses more clearly than any earlier writer the idea of substances that are both original and simple. True, he does not use what became the technical term for element in Greek, *stoicheion*, which is not introduced until Plato, but he refers to earth, water, air and fire as *rhizomata*, 'roots', in a well-defined sense. First, the roots themselves do not come to be, but are eternal and uncreated: they are, then, elemental in the sense of original substances. And secondly, they—together with Love and Strife which are responsible for mixing and separating them—are what everything else in the world is made of. Above all, Empedocles draws a clear distinction between compounds and what they are composed of. In fragment 23, for instance, he compares the variety of different things which come to be from the roots with the variety of colours which a painter can make from his pigments, and he concludes that fragment by insisting that the roots are the source of every other kind of substance:

> Do not let error overcome your mind that there is any other source than this [the four roots] for all the mortal things that appear in their countless numbers.

The idea of a constituent element is more definitely grasped by Empedocles than by any earlier Presocratic philosopher. His roots are both eternal and simple—they are the irreducible constituents into which other things can be analysed. Nevertheless his conception of elements, like that of every other Greek scientist, differs from the modern notion in at least one obvious but vital respect: they are not chemically pure substances. Empedocles held that things are made of earth, water, air and fire, but 'earth' is a term which was applied to a wide variety of solid substances, 'water' was generally used not only of various liquids, but also of the metals (because they are fusible), and 'air' was the Greek term for any gas. We should not, then, think of Empedocles' roots as pure substances like the oxygen and hydrogen of post-Lavoisier chemistry. On the other hand this may help us to understand Empedocles' actual choice of elements, which is not quite as arbitrary as might at first appear. Whatever other factors influenced his theory, earth, water and air represent, very approximately, matter in its solid, liquid and gaseous states: and fire, thought of as a substance rather than as a process, was naturally included as a fourth 'element' on a par with the other three.

Empedocles' second important contribution to the development of physical theory is in his use of the idea of proportion. We have seen that he postulated four roots and made all other substances compounds of them. But in answer to the difficult question of how a finite number of roots can give rise to an apparently almost unlimited number of different substances, he made what can only be described as an inspired guess. He said that different substances are formed by the roots combining

41

in different proportions, and he evidently assumed that any particular substance is always formed by the roots combining in a fixed and definite proportion.

Two comments must be made on this theory. First, the idea of proportion was one that had already been used extensively by the Pythagoreans in their musical theory, in their cosmology and in their ethics. And the idea undoubtedly had ethical associations for Empedocles too. He uses *Harmonia* as a synonym for *Philia*, Love, and both in his cosmology and more especially in his religious poem, the *Purifications*, Love is generally conceived as a good principle, bringing about good results, while Strife is described as evil and accursed.

Secondly, the extent to which he worked out the detailed application of his theory to specific substances is very limited. Only two of the extant fragments refer to the proportions of the roots in different compounds, fragment 96 which suggests that bone was composed of fire, water and earth in the ratio 4:2:2, and fragment 98 which indicates that blood and different kinds of flesh were made of the four roots in equal proportions—there being a special reason for this, as blood is the seat of cognition and apprehends the elements on the principle of 'like-to-like'. Even when we bear in mind the fragmentary nature of our evidence, it seems likely that Empedocles made concrete suggestions about the composition of only a very few substances, and he clearly made no attempt to follow up such suggestions as he did make by conducting tests on different substances to throw light on their constitution. Having seen that a vast variety of substances could *theoretically* be accounted for by supposing that the roots combine in different proportions, he left it at that, and at no stage entertained the idea of inquiring into the problem further in a programme of empirical investigations. Yet when all this has been said, the basic contribution that his idea makes to chemical theory remains. The Law of Fixed Proportions states that chemical compounds contain their constituent elements in fixed and invariable proportions by weight, but long before the law in that form was established by experiment, Empedocles had arrived by conjecture at a similar general principle.

About the same time as Empedocles another solution to Parmenides' denial of change was put forward by a philosopher of a very different kind, Anaxagoras of Clazomenae. Whereas Empedocles, like Parmenides himself, came from the western confines of the Greek world, Anaxagoras was born in Ionia and lived much of his life in Athens, where he was the friend and teacher of Pericles; and whereas Empedocles, like Parmenides again, wrote in verse, Anaxagoras followed the Ionian tradition, set by Anaximander, Anaximenes and Heraclitus, and chose prose as his medium. Whereas Empedocles composed not only a work on nature, but also a religious poem which owes much to Pythagorean beliefs, Anaxagoras' interests were entirely in natural philosophy, and he was prosecuted by the Athenians on a charge of impiety, although the motives of his accusers were partly, or mainly, to discredit Pericles politically through Anaxagoras. The question of whether either philosopher knew the other is an open one. But despite this, and despite the obvious temperamental differences between the two men, their answers to the challenge of Parmenides' philosophy had a good deal in common.

Like Empedocles, Anaxagoras tackled the problem of the foundations of knowledge. Fragment 21 makes a by now conventional reference to the weakness of sense-perception, but fragment 21a is more important and original, for in it Anaxagoras states the principle that 'phenomena' provide a 'vision' of 'things that are obscure'—that is, that the evidence of the senses provides the basis for inferences concerning what cannot be directly observed. Then like Empedocles again, Anaxagoras resolved the chief problem left by Parmenides by denying the uniqueness of what exists, while retaining the principle that nothing can come to be from not-being. In fragment 17 he says that nothing comes to be or perishes, but is mixed and separated from existing

43

things. This was, as we noted, Empedocles' position too: yet what Anaxagoras meant by 'existing things' is quite different from what Empedocles would have meant by the corresponding phrase. Whereas for Empedocles the existing things that mix and are separated are the four roots, in Anaxagoras they include every kind of natural substance, and not only such things as hair and flesh and gold and stone, but also the opposites—'the hot', 'the cold', 'the wet', 'the dry' and so on—these being conceived as things rather than as mere qualities.

One of the problems that particularly engaged his attention was that of nutrition and growth. Aristotle reports that he asked how blood or flesh, for example, can come to be, and one of our later sources gives what purports to be a verbatim quotation (fragment 10): 'How could hair come to be from not-hair, or flesh from not-flesh?' The implication is that 'hair', 'flesh' and so on must already exist in some form in our food. By the same argument, wood, leaves and the different sorts of fruit must pre-exist in the earth and water that are the nourishment of plants. Indeed Anaxagoras states this doctrine in its most general form: 'In *everything* there is a portion of *everything*.' And since at no stage can hair, for instance, come to be from not-hair, it is clear that hair, flesh and so on must have existed from the beginning, in the original mixture of all things.

In the beginning, as Anaxagoras puts it (fragment 1), 'everything was together.' And now too 'all things have a share in all things.' What we recognise as a lump of gold is predominantly gold, but contains a small proportion of every other substance as well. What we recognise as wheat contains flesh, bone, blood, and not only them, but also gold, iron, stone and every other kind of natural substance. When we digest the wheat, some of the flesh and bone and blood it contains separates out and joins the flesh, bone and blood in our bodies, but this process of separation is never completed, since there *remains* a portion of everything in everything.

Where Empedocles suggested that four elements are

44

sufficient to account for all known substances, for the four roots combine in different proportions to form different compounds, the gist of Anaxagoras' argument is that no natural substance is more elemental, in the sense of simple, than any other. Every kind of natural substance existed in the primordial mixture when everything was together: and every type of natural substance exists today in every object we see around us. This seems, no doubt, a highly uneconomical theory, and so it is if we consider the number of *substances* whose existence Anaxagoras postulates in any given object and in the world in general. Yet from another point of view the theory is, on the contrary, extremely economical, that is in the number of *assumptions* it employs. It is an attempt to resolve a wide variety of problems associated with change by using the single principle 'in everything a portion of everything'.

Both Empedocles and Anaxagoras proposed ingenious and original physical theories. But the most famous and most influential of the fifth-century systems was the atomic theory first suggested by Leucippus of Miletus and then developed by Democritus of Abdera. This is rightly considered the culmination of Presocratic speculation. The problem of assessing it fairly, however, has been aggravated by the tendency to assimilate ancient atomism to modern theories that bear the same name, even though there are fundamental differences both in the content of the theories themselves and in the grounds on which they were advanced. Dalton's theory, for instance, differs from ancient atomism in allowing different elemental substances, and, since the analysis and splitting of the atom, modern 'atomic' theory is not an atomic theory at all in the Greek sense, for the word *atomon* in Greek *means* indivisible.

The basic postulate of ancient atomism in its original, fifth-century, form was that atoms and the void alone are real. The differences between physical objects, including both qualitative differences and what we think of as differences in substance, were all explained in terms of modifications in the shape, arrangement and position of the atoms. The examples that Aristotle gives to illustrate these three modes of difference between the atoms are A and N (shape), AN and NA (arrangement) and Ⅱ and H (position).

The atoms are infinite in number and dispersed through an infinite void. They are, moreover, in continuous motion, and their movements give rise to constant collisions between them. The effects of such collisions are two-fold. Either the atoms rebound from one another, or if the colliding atoms are hooked or barbed or their shapes otherwise correspond to one another, they cohere and thus form compound bodies. Change of all sorts is accordingly interpreted in terms of the combination and separation of atoms. The compounds thus formed possess various sensible qualities, such as colour, taste, temperature and so on, but the atoms themselves remain unaltered in substance.

Just as much as the other systems we have considered, this theory was evolved as a response to the problem set by Parmenides and the other Eleatic philosophers. Indeed in postulating a single elemental substance, Leucippus remained closer to Parmenides' own conception than either Empedocles or Anaxagoras. Like the one unchanging being of the Way of Truth, each individual atom is ungenerated and indestructible, unalterable, homogeneous, solid and indivisible. Leucippus may be said to have postulated an infinite plurality of Eleatic ones, and he may even have been directly influenced by an argument that Melissus had put forward when he tried to show that 'if there were many, they would have to be as the one is' (fragment 8): although Melissus had evidently intended this argument to suggest the absurdity of the notion of a plurality, Leucippus believed that to postulate just such a plurality of objects like the Eleatic one resolved the problem of change. Leucippus also agreed with the Eleatics that without void movement is impossible. Yet whereas the Eleatics denied the existence of the void, Leucippus maintained that not only 'being' or 'what is'—the atoms—but also 'not-being' or 'what is not'—the void—must be considered real. This was the key step by which he reinstated both plurality and change. The void is that which separates the atoms and that through which they move.

The main features of the theory the atomists proposed in answer to the Eleatics are clear. But how far did either Leucippus or Democritus go in working out their theory or applying it in detail? Once again we must make allowances for the fragmentary nature of our information, much of which derives from sources hostile to atomism. But even so it seems likely first that the atomists left certain conceptual difficulties in their theory unresolved, and secondly that they were sparing in their attempts to apply it to explain specific phenomena.

It is not clear, for example, whether they considered their atoms mathematically, as well as physically, indivisible. The atoms certainly could not be split physically, but did the atomists also conceive them as logically or mathematically indivisible, that is as having no parts? We cannot be sure of the answer to this question, but certain texts in Aristotle (for example *On Coming-to-be and Passing-away* 315 b 28 *ff*) seem to suggest that they drew no distinction between the limits of physical, and those of mathematical, divisibility. Unless Aristotle is grossly misrepresenting them, they seem not to have recognised that if the atoms differ in shape, this implies that they have parts and so must be considered mathematically divisible.

Again some of our sources suggest that the variety in both the shapes and the sizes of the atoms is infinite. The opponents of atomism exploited this assumption to raise objections to the theory, such as the absurdity of an atom the size of the world. But while we can be sure that this was a conclusion that Leucippus and Democritus would have resisted, we cannot be certain what defence they would have offered, or whether indeed the difficulty had occurred to them.

Leucippus was undoubtedly responsible for the foundations of the atomic theory, but there is little evidence that he made any attempt to apply the doctrine in detail to explain natural phenomena. His position seems to have been similar to that of Empedocles who, once he had seen how a vast variety of compounds might in principle be accounted for, made only a few concrete suggestions concerning the constitution of specific substances. Democritus' interests, on the other hand, were very wide. Though few fragments of his work survive, the titles of his books indicate the range of his inquiries: apart from physics and cosmology, he wrote on astronomy, zoology, botany and medicine, besides composing treatises on a number of technical subjects, such as agriculture, painting and warfare. Moreover he applied the atomic theory in detail in at least one area. This is his doctrine of sensible qualities.

In his theory of knowledge he describes the knowledge provided by the senses as 'bastard' knowledge, contrasting it with the 'legitimate' knowledge of the mind, although he recognised that the mind derives its data from the senses. What the senses perceive are the secondary qualities which are due to the differences in the shapes, sizes and arrangements of the atoms, but atoms and the void alone are real, and these secondary qualities exist 'by convention' only. Not content merely to propose this general theory, he put forward detailed doctrines—reported and criticised at length by Theophrastus in his *On the Senses*—in which he related specific tastes, colours, smells and so on to specific atomic configurations. Thus an acid taste is composed of angular, small, thin atoms, and a sweet taste of round, moderate-sized ones. Again he associated what he took to be the four primary colours, black, white, red and yellow, with certain shapes and arrangements of atoms, interpreting the other colours as compounds of these four. This is the first attempt to give a detailed account of the physical basis of sensation. We may, however, observe that for all the ingenuity of the atomic theory, when it came to applying it in detail Democritus fell back on fairly crude physical analogies, in which 'sharp' tastes, for instance, are associated with figures with 'sharp' angles.

In sum, the main preoccupation of the later Presocratic philosophers was with the problem of change. It is true that they advanced explanations of many different phenomena in meteorology, geology, physiology, embryology and other fields. Thus such problems as why the sea is salt, or why the Nile floods, were much discussed in the fifth century. Empedocles was one of those who attempted an account of the process of respiration in which he compared the way the breath is drawn into and expelled from the body with the action of a clepsydra, an instrument for lifting liquids. And several theorists debated the cause of sex differentiation in the embryo.

Nevertheless the key problem in natural philosophy was a general one, the nature of coming-to-be and change. The answers proposed took the form of a series of physical theories, that is, of accounts of the ultimate constituents of matter. But the problem was in origin a philosophical one, which was posed in its sharpest form by Parmenides' denial of the possibility of change, and each of the fifth-century theorists appreciated that to deal with Parmenides' problem it was necessary also to tackle the question of the foundation of knowledge. Physics was, indeed, inextricably bound up with epistemology in the fifth century. Empedocles, Anaxagoras, Leucippus and Democritus were chiefly engaged not in programmes of research, but in discussions of a highly abstract nature in which what counted was not the empirical data that could be adduced in support of a theory, so much as the economy and consistency of the arguments on which it was based.

5

The Hippocratic Writers

So far we have been dealing with the work of men who are usually thought of chiefly as philosophers, but it is now time to turn to our other main source of information concerning early Greek natural science, the texts of the Hippocratic Corpus. Here we do not have to rely on the isolated fragments of statements that happen to have been preserved because they are quoted by later writers, since more than fifty entire treatises have come down to us. Although the Corpus as a whole came to be named after the great fifth-century physician Hippocrates, it is nowadays thought unlikely that he wrote any of the treatises himself. The collection probably derives, in the main, from the library of a medical school, but the works it contains vary a great deal in both date and style. While most of them come from the late fifth or the fourth century, some are more recent still. Besides textbooks on particular branches of medicine, such as surgery or gynaecology or dietetics, they include records of day-to-day clinical practice, commonplace books and lectures addressed to a general public on such topics as the constitution of man.

Not all the authors were themselves medical practitioners. Some of the public lectures were composed by men who had little or no clinical experience, who were teachers, rather than doctors, by profession. In the second half of the fifth century the demand for knowledge of all sorts grew very rapidly in Greece, and we meet a new type of professional teacher—the sophist—who taught for a fee and travelled about the Greek world to do so. Some of these sophists lectured on a wide variety of topics—Hippias of Elis, we are told, was prepared to teach any science or art—and even such a technical subject as medicine was often discussed in public debate by men who had never practised as doctors. In the fourth century, too, Aristotle distinguished three types of person who had a claim to speak on medical matters, the ordinary practitioner (*demiourgos*), the master of the craft (*architektonikos*) and the man who has studied medicine as part of his general education.

Although we may speak loosely of those who engaged in medical practice full-time as professional doctors, medicine was not a profession in the fullest modern sense of that term. Since doctors possessed no legally recognised professional qualifications, anyone could claim to heal the sick.[1] Yet we find the author of the treatise *On Ancient Medicine*, for example, insisting on the distinction between the doctor who is experienced in the 'art' and the mere layman (*idiotes*), and the treatise *On the Sacred Disease* is equally emphatic concerning the difference between the true representative of the medical art and the quack or charlatan. In general, technical knowledge and skill were transmitted, in medicine as in other arts and crafts, by means of a system like that of apprenticeship, where the young— often, but not exclusively, the sons of doctors—were taught by already established practitioners. Already in the late sixth century certain city-states, such as Croton and Cyrene, were famous for their doctors, and in the fifth century both Cos (the birth-place of Hippocrates) and Cnidus, especially, developed flourishing medical schools in both senses of the word 'school': they became the main centres for the teaching of medicine, and the doctors associated with either place shared certain medical doctrines and practices.

As the doctor had no formally recognised professional status, the conditions of medical practice were precarious. Occasionally we hear of a doctor being employed, usually for a year at a time, by a city-state, although the duties of such public physicians are far

[1] Thus the distinction between the doctor and the gymnastic trainer, in particular, was often a fine one.

from clear, and it may be that all that the state required, and paid for, was that the doctor should reside and practise in the city. But in general the Greek doctors practised privately and travelled from city to city in response to the varying demand for their services. Thus the work called *On Airs, Waters, Places* is designed to help the itinerant physician to anticipate the types of diseases that are likely to occur in towns with different climates and locations.

Unless he already possessed an established reputation, the itinerant doctor was faced with a recurrent problem in having to build up a clientèle in each city he stayed in. Here the practice of 'prognosis', which included not only foretelling the outcome of a disease, but also describing its past history, was an important psychological weapon. As the author of the treatise *Prognostic* puts it (Chapter 1):

> If he is able to tell his patients when he visits them not only about their past and present symptoms, but also to tell them what is going to happen, as well as to fill in the details they have omitted, he will increase his reputation as a medical practitioner and people will have no qualms in putting themselves under his care.[1]

But if it was difficult enough to persuade his patients that he understood their condition, it was usually far harder to attempt a cure with the limited means at his disposal. The methods of treatment mentioned in the Hippocratic Corpus consist of a very few general types, of which the most important were surgery, cautery, blood-letting, the administration of purgative drugs, and, especially, the control of 'regimen', that is diet and exercise. The role of the doctor was, therefore, often a defensive one. He tried first and foremost to preserve health, generally conceived as a balance between opposites, and when this balance was upset, he largely relied

[1] From the translation in *The Medical Works of Hippocrates* by J. Chadwick and W. N. Mann (Oxford, Blackwell, 1950).

on nature to effect its own cure, concentrating his efforts to help, or at least not to hinder, this process.

But despite the hazards of medical practice, some doctors were highly successful and earned large sums of money. This can be seen, for example, from the story of Democedes of Croton, reported in Herodotus (III, 131). Democedes was employed in three successive years by Aegina, Athens and Polycrates of Samos, and on each occasion was offered a higher salary, first one talent, then 100 minae, and then two talents.[1] However, such a case was no doubt exceptional, and the income of the average doctor must obviously have varied within wide limits according to his skill and reputation. It is noteworthy that the treatise entitled *Precepts* warns doctors not to be too eager to discuss fees with a sick patient, as this may cause him undue anxiety, and the writer also recommends that in fixing the fee the doctor should take the patient's means into account and even be prepared to give treatment for nothing. Although this treatise is a late, post-Aristotelian work, a similar code of behaviour may go back to the fifth century. We can at least be sure that people from all walks of life were treated by some of the best doctors in our period. The patients whose case-histories are recorded in the treatises known as the *Epidemics* include men and women, rich and poor, citizens and visitors from abroad, free men and slaves. Although Plato spoke in the *Laws* (720cd) of free men being treated by freeborn doctors and slaves being treated by slaves, such a distinction runs quite contrary to the practice of the doctors represented in the *Epidemics*.

The training, interests and whole manner of life of those who practised medicine were, in many respects, quite different from those of the philosophers.[2] Yet in

---

[1] There were sixty minae to the talent, and a hundred drachmae to the mina. Some indication of the real values of these units can be gathered from the fact that the normal daily wage of a skilled worker in the late sixth and early fifth centuries was a drachma.

[2] See further below, Chapter 9.

some ways the contributions that the doctors made to
natural science may be said to parallel the work of the
Presocratic cosmologists. Thus one of the chief battles
that the Hippocratic physicians had to fight was to get
it accepted that disease is a natural phenomenon, the
effect of natural causes. Just as the Milesians had rejected the idea of divine interference in such domains as
meteorology and astronomy, so too the doctors did in
medicine. The treatise *On the Sacred Disease*, in particular, provides valuable evidence illustrating how one
writer refuted superstitious beliefs, in this case concerning epilepsy.

The work begins:

> I do not believe that the 'Sacred Disease' is any more
> divine or sacred than any other disease but, on the con-
> trary, has specific characteristics and a definite cause.
> Nevertheless, because it is completely different from other
> diseases, it has been regarded as a divine visitation by those
> who, being only human, view it with ignorance and aston-
> ishment.[1]

He compares those who first thought the disease sacred
with the 'magicians, purifiers, quacks and charlatans' of
his own day. First he accuses them of dishonesty. Calling
the disease sacred is just a cover for their own ignorance.
They prescribe a method of treatment, including the use
of charms and the prohibition of certain types of food
and clothing, but this is simply so that if the patient
recovers, they get the credit, while if he dies, they have
a cast-iron excuse—that the gods are to blame. He
argues, too, that it is inconsistent, if one believes the
disease to be divine, to try to cure it by such facile
methods as purifications and charms, and he rejects as
positively impious the idea that the gods are responsible
for defiling a man's body.

Above all the writer suggests (Chapter 5) that the

[1] From the translation in *The Medical Works of Hippocrates* by
J. Chadwick and W. N. Mann (Oxford, Blackwell, 1950).

disease cannot be any more sacred than any other since it attacks only those of a phlegmatic disposition:

> Another important proof that this disease is no more divine than any other lies in the fact that the phlegmatic are constitutionally liable to it while the bilious escape. If its origin were divine, all types would be affected alike without this particular distinction.

He explains the disease as arising from a discharge from the brain and to support this theory he suggests examining an animal, such as a goat, that has suffered from the disease (Chapter 14):

> If you cut open the head, you will find that the brain is wet, full of fluid and foul-smelling, convincing proof that disease and not the deity is harming the body.[1]

Although traces of superstition persist in some of the Hippocratic writings, the vast majority of the doctors there represented agreed with the author of *On the Sacred Disease* in rejecting the idea that diseases are caused by supernatural agencies. The appeal to what can be found out by looking—in this case by conducting a post-mortem examination on a goat—takes us to an even more important feature of Hippocratic medicine, namely the appreciation of the value of carrying out detailed and methodical observations in the diagnosis of disease. One treatise that pays special attention to the points the doctor should look for in examining his patients is *Prognostic*, where the writer is particularly concerned with 'acute' diseases, that is those accompanied by high fever such as pneumonia or malaria.

[1] One may compare the story that Plutarch (*Pericles*, Chapter 6) tells to illustrate how Anaxagoras helped to emancipate Pericles from superstitious beliefs. A one-horned ram was brought to Pericles, and the seer Lampon said that this foretold that Pericles would become supreme ruler. Anaxagoras, however, had the skull opened and thereupon explained the phenomenon as due to natural causes. While those present admired this demonstration at the time, according to Plutarch, later on it was Lampon that won their admiration, when his prophecy was fulfilled.

First, he should examine the patient's face, for example the colour and texture of the skin. The eyes are especially important:

> For if they avoid the glare of light, or lacrimate without due cause . . . or if the whites are livid or show the presence of tiny dark veins, or if bleariness appear around the eyes, or if the eyes wander, or project, or are deeply sunken . . . ; then all these things must be considered bad signs and indicative of death (Chapter 2).[1]

The doctor should also inquire about how the patient has slept, about his bowels and his appetite; he should take into account the patient's posture, his breathing and the temperature of the head, hands and feet, and separate chapters are devoted to how to interpret the symptoms to be found in the patient's stool, urine, vomit and sputum.

Prognostic lays down some excellent general principles for the examination of patients, but to see how far Greek doctors put these principles into practice we must refer to the *Epidemics*, especially books I and III. These contain first what are known as the 'constitutions'—general descriptions of the climatic conditions accompanying certain outbreaks of diseases—and then a number of detailed case-histories in which the daily progress of a particular patient's illness is described. The entries under each day vary from a single remark to a lengthy account, and in some cases occasional observations continue to be recorded up to the 120th day from the onset of the disease. Case three of the second series in *Epidemics* III, for example, begins:

> At Thasos, Pythion, who lay beyond the temple of Heracles, had a violent rigor and high fever as the result of strain, exhaustion and insufficient attention to his diet. Tongue parched, he was thirsty and bilious and did not sleep. Urine rather dark, containing suspended matter which did not settle.

[1] From the translation in *The Medical Works of Hippocrates* by J. Chadwick and W. N. Mann (Oxford, Blackwell, 1950).

Second day: about midday, chilling of the extremities, particularly about the hands and head, showed both aphasia and aphonia, and he also was dyspnoeic for a long time. Then he became warm again and thirsty. A quiet night; slight sweating about the head.

Third day: quiet. Late in the day, about sunset, slight chilling, nausea, disturbed bowels followed by an uneasy sleepless night. Passed a small, constipated stool.

Fourth day: morning quiet. About midday, all symptoms more pronounced; chilled; aphasia, aphonia became worse. After a while he became warm again and passed dark urine containing suspended matter. A quiet night; slept.[1]

And the patient's condition continues to be recorded daily until his death on the tenth day.

There is nothing to equal these case-histories in extant medical literature until the sixteenth century—and one of the chief figures responsible for the revival of detailed clinical histories, Guillaume de Baillou (born c. 1538), explicitly took the Hippocratic *Epidemics* as his model. Each case-history is a bare, methodical account of the symptoms with the minimum of interpretative comment. Reference is seldom made to the treatment prescribed, and, as our case shows, the writer had no compunction in admitting his failure to effect a cure. Of the forty-two cases described in these two books, twenty-five, or nearly 60 per cent, ended in death—which contrasts with the reluctance to refer to unsuccessful cases that is shown by clinicians in later centuries.

The primary aim of these books of the *Epidemics* is to provide as exact a record as possible of the cases investigated. But while the writer proposed no over-all theory of disease, many of the terms he uses are 'theory-laden' and reveal his assumptions concerning the nature and causes of diseases. Thus although he presents no schematic doctrine of humours, such as we find in other treatises, he often refers to 'bilious' and 'phlegmatic'

[1] From the translation in *The Medical Works of Hippocrates* by J. Chadwick and W. N. Mann (Oxford, Blackwell, 1950).

matter in the patients' discharges. He believed that the course of 'acute' diseases is determined by what were known as 'critical days', when marked changes take place in the patient's symptoms, and he adopts the common Greek classification of fevers into 'tertians', 'quartans' and so on according to the length of the cycle which was discovered or imagined in these changes. Indeed this doctrine of critical days must be recognised as one of the main motives the writer had for carrying out and recording his observations day by day. *Epidemics* I and III provided an important model of method, demonstrating what it was to conduct systematic observations in the field of clinical medicine. But these observations were guided by, and they reflect, the writer's theoretical assumptions and interests.

Naturally enough the principal issue that preoccupied many of the Hippocratic authors is the general question of the causes of diseases. The variety of views that were maintained is extraordinary, equalling or even surpassing the variety of physical theories found in contemporary philosophy, and providing the subject of as intense and protracted a debate. They range all the way from those who argued that all diseases have a single origin, to those who held that there are as many different diseases as there are patients—that wherever there is any difference between two sets of symptoms, two different diseases must be diagnosed.

Some of the treatises that tackle this problem are exhibition pieces designed for rhetorical performances. Thus the author of *On Breaths* runs through a number of common diseases pointing out that 'breath' or 'air' has some connection with each of them, and then triumphantly declares that he has demonstrated that this is the cause of all diseases. But the discussion was not always conducted at such a low level. Even if attempts to state the origins of diseases in general were often too oversimplified to be of much value, several writers have sensible points to make on, for example, the concept of a cause. Thus the author of *On Ancient Medicine*

58

(Chapter 21) warns against what he describes as a fault common among doctors and laymen alike—they confuse what is merely coincidental to the disease with its cause. If a patient has done anything unusual, such as eating strange food, just before he becomes sick, they jump to the conclusion that this must be the cause of the disease. And *On Regimen in Acute Diseases* (Chapter 11, Littré) points out that the same symptoms may have very different explanations and issues a similar complaint against medical practitioners who ignore this.

The problem of the causes of diseases was closely connected with the question of the constituent elements of the human body. Here the interests of the doctors overlapped those of the natural philosophers, and this sparked off a controversy between them on the right way of studying the problem. The writer of *On Ancient Medicine*, especially, protested vigorously against those who imported the methods of the philosophers into medicine. In particular he condemned those who based their theories on what he calls *hypotheseis*, postulates or assumptions, such as 'the hot', 'the cold', 'the dry' or 'the wet'. The first criticism he makes of these theories is that they 'narrow down the principle of the causes of diseases'. Medicine, he says in Chapter 1, is an art, *techne*, which has both good and bad practitioners. The treatment of the sick is not a matter of chance, but requires skill and experience:

> So I considered that medicine has no need of an empty postulate, like obscure and problematic subjects, concerning which anyone who attempts to hold forth at all is forced to use a postulate, as for example about things in heaven or things under the earth: for if anyone were to discover and declare the nature of these things, it would not be clear either to the speaker himself or to his audience whether what was said was true or not, since there is no criterion to which one should refer to obtain clear knowledge.

Here the writer distinguishes between different

59

inquiries not merely by their subject-matter, but also by their method. He differentiates between those inquiries where some sort of postulate is necessary, and those like medicine, as he believed, where it is not. But when he refers to what we should call astronomy, meteorology and geology as examples of inquiries where a postulate is necessary, he should not be interpreted as approving of the use of postulates in these subjects at least. On the contrary, the very fact that they have to make use of a postulate is enough, in his eyes, to condemn those inquiries as worthless, for this is what he implies in the remark that 'it would not be clear either to the speaker himself or to his audience whether what was said was true or not, since there is no criterion to which one should refer to obtain clear knowledge'. This is, in effect, a statement of the need for scientific theories to be testable. Speculation about what goes on in the sky or under the earth is worthless because unverifiable, at least by the writer's standards of verifiability.

The writer of *On Ancient Medicine* makes some important methodological recommendations, but it is right to consider how far he carried them out in his own medical and biological doctrines. His theories of the origin of diseases and of the constituent elements of the body are, admittedly, more complex than those based on 'the hot', 'the cold' and so on, which he had singled out for particular criticism. In Chapter 14, for instance, he says that there are many different things, with a great variety of different 'powers' or effects, in the body. Yet the examples he gives are such things as the salty, the bitter, the sweet, the acidic, the astringent and so on. While this increases the number of component substances in the body, his theory is almost as arbitrary as those based on the hot and the cold. In Chapter 13 *ff*, for example, he had questioned how those opposites were applied in practice: 'When a baker bakes bread, is it the hot or the cold or the dry or the wet that he takes away from the wheat?' Yet one might ask a similar question in connection with the writer's own theory. How the salty and the astringent are to be defined and identified in practice raises equally difficult problems. The principle implied in Chapter 1 is that theories should not be based on arbitrary assumptions and they should be capable of verification. But so far as such general problems as the origin of diseases or the elements in the body are concerned, this was, at this stage, a largely impracticable ideal.

This last point can be further illustrated by considering another medical writer who displays an awareness of contemporary philosophical speculation, the author of *On the Nature of Man*. Chapter 1 begins with the declaration that the writer will not discuss 'the nature of man' beyond its relevance to medicine. He criticises those who claimed that man is air or fire or water or earth, each of whom argued that man is a unity and each of whom 'adds to his account evidences and proofs which amount to nothing'. To discover how ignorant they are, he says, it is enough to go along to their debates, where the victory goes to whoever happens to have the glibbest tongue in front of the crowd: 'such men seem to me to undo themselves by their ignorance . . . and to establish the theory of Melissus'—that is the view that the one is unchanging.

The writer confidently claims that he will declare the constituents of man, and these turn out, in Chapter 4, to be the four humours, blood, yellow bile, black bile and phlegm, each of which he associates with two of the four primary opposites, hot, cold, wet and dry, and each of which predominates in the body in one of the four seasons, spring, summer, autumn and winter. But having been so scathing about the lack of evidence adduced by his opponents for their theories based on a single element, how far does he succeed in fulfilling his promise to demonstrate his own highly schematic four-fold doctrine? The main evidence he brings relates to the effects of certain drugs that were used to draw out, or purge, phlegm and the two sorts of bile. But while the evidence he cites indicates the presence of certain substances in the body well enough, it fails to establish, or even to be relevant to, what he claims to prove, namely that they are congenital in the body.

On Ancient Medicine and *On the Nature of Man* both resist the invasion of medicine by philosophical ideas and methods, and in doing so both writers adopt what is, broadly speaking, an empiricist position. They reject arbitrary speculation and stress the need to bring evidence to support any general theory that is proposed. The interest of these treatises lies partly in that they indicate a growing awareness of problems of method, and the theses they advance do indeed reflect an important difference in approach between the doctors and the philosophers. It was above all in dealing with the sick that the doctors learned the importance of assembling evidence and the need for caution in attempting causal theories. But on many general theoretical issues in pathology and physiology the ultra-cautious empiricism advocated by *On Ancient Medicine* was a council of unattainable perfection, for to have done away with 'unverifiable' postulates would have been to abandon theorising altogether.

My last two examples of problems discussed by the Hippocratic authors again illustrate the methods the doctors used, and bear on the relation between philosophy and medicine. These are the problems of growth and reproduction. Growth was generally explained, as for example in *On the Nature of the Child*, on the principle of 'like-to-like': each of the constituent substances in the body draws to itself the same substance from the food and drink we consume. This principle was so common and was applied to such a wide range of problems in antiquity that we should not try to specify a particular source by whom the Hippocratic writers might have been influenced. The Greeks themselves connected this idea with proverbial sayings expressing the notion that 'birds of a feather flock together', and it was used to explain cognition and growth by several of the Presocratic philosophers.

But associated with the problem of growth were those of generation and reproduction and here more direct influences were sometimes at work. Among the many questions raised was that of the composition of the seed. The problem might be put: how do all the different substances in the mature animal or plant arise from the apparently homogeneous seed? The doctrine that the seed is drawn from every part of the body and contains each of the substances in it was probably put forward first by the atomist Democritus. It is implied or assumed in several Hippocratic works and stated in the group of embryological treatises that contains *On Generation* and *On Diseases* IV. Thus *On Generation* puts it (Chapter 3):

> I say that the seed is separated off from the whole of the body, both from the hard parts and from the soft, and from all the fluid in the body, and there are four sorts of fluid, blood, bile, water and phlegm.

We have seen that several doctors objected to the importation of the ideas of the philosophers into medicine, and yet the philosophers were responsible for many important biological doctrines. It so happens that on the problem of the composition of the seed most of the main ideas that were put forward in the late fifth and fourth centuries were supplied by men who were not themselves medical practitioners, notably first by Democritus and then later by Aristotle.

On many obscure physiological and embryological problems the arguments adduced by the philosophers and medical writers alike were largely, and one may add often inevitably, abstract and dialectical in character. Such questions as the composition of the seed, the transmission of inherited characteristics and the origin of sex differentiation were not issues that could be settled by direct appeals to easily accessible evidence. That is not to say, however, that no attempts were made to bring empirical methods to bear on the study of embryology. The same group of works that adopts Democritus' theory of the seed, and contains many other much wilder embryological doctrines, also contains our first extant reference to a systematic investigation of the growth of a hen's egg. In *On the Nature of the Child* (Chapter 29) the writer suggests incubating a batch of twenty hen's eggs and opening one each day in order to observe the embryo at the various stages of its development.

In the text just cited, it is not clear whether the writer merely opened the egg or whether he also dissected the embryo, and the extent to which the method of dissection was known and practised during the fifth and early fourth century is problematic. Passages in Aristotle establish that some of his predecessors and contemporaries undertook dissections, and a late source, Chalcidius (c. 400 A.D. according to the most recent editor), even suggests that the method goes back to Alcmaeon of Croton, a fifth-century theorist who combined medical and cosmological interests. But direct references to the use of the method are very rare in the Hippocratic Corpus, the main exception being the brilliant but late (post-Aristotelian) work *On the Heart*. It appears, then, that the technique was not used at all extensively until Aristotle himself in the mid fourth century, and dissections were not carried out on human bodies until even later, in Alexandria in the mid third century. Such progress as the Hippocratic writers achieved in the field of anatomy was the result not of deliberate research, so much as of their clinical experience, particularly in surgery. While most of these authors evidently had only the vaguest ideas of internal human anatomy, such surgical treatises as *On Fractures* and *On Joints* show some working knowledge of the bone structure, at least, a knowledge that was no doubt gained from having to deal with fractures, dislocations and wounds of different types.

The life of the medical man of the fifth or fourth century was, as I noted at the beginning of this chapter, in many ways very different from that of the philosopher. On certain theoretical problems, such as the constituent elements of man, or generation and growth, the interests of the two types of writer overlapped, and in criticising the cosmologists' views some of the doctors were led to raise far-reaching questions concerning the relation between medicine and philosophy and the correct way of tackling the problems in question. One of the results of this dispute was to draw attention to questions of method, even if in practice the theories the Hippocratic writers advanced often have much more in common with those they rejected than one might expect from the nature of their objections. Nevertheless one fundamental point of difference does separate most of the medical writers from the philosophers, and this lies not in the type of theory they presented, nor in the method they adopted, but in the underlying motive for which they undertook the inquiry. I shall be returning to discuss this more fully in my final chapter; it is enough for the moment to observe that unlike the philosophers, the doctors had, in the long run, a practical end in view. As the writer of *On Ancient Medicine* puts it, medicine is an art with both good and bad practitioners. The ultimate aim that the doctors had in mind was, in fact, the treatment of the sick.

6

Plato

THREE changes that profoundly affected the subsequent development of Greek thought took place in the second half of the fifth century. First, I have already noted the expansion of education that is associated with the sophistic movement. Whereas the traditional Greek education was confined to grammar, music and poetry, the sophists were prepared to lecture on any subject that anyone would pay them to teach. Secondly, in Cicero's famous phrase (*Tusculan Disputations* V, 4, 10), Socrates 'called down philosophy from the skies'. Whereas earlier philosophers had devoted more attention to physics and cosmology than to ethics, the reverse was true not only of Socrates himself, but also of many of the sophists. Thirdly, Athens became the chief intellectual centre of Greece. Whereas most earlier philosophers lived and worked either in Ionia or in Magna Graecia, from the generation of Socrates onwards more and more of the important thinkers either were born at Athens or spent a considerable part of their lives there, and in the fourth century this development was accentuated when first Plato and then Aristotle founded schools—the Academy and the Lyceum—that attracted philosophers and scientists from all over Greece.[1]

Socrates himself is rightly taken as marking a turning-point in Greek thought, but his importance, and that of such sophists as Protagoras, lies in the field of moral philosophy rather than in science. As a pupil of Socrates, Plato shared his master's passionate concern with moral issues, but, unlike Socrates, Plato was also a figure of great importance in the development of Greek science. This was not merely because he founded the Academy[2]

[1] See pp. 127 *ff.*
[2] The date of its foundation cannot be determined precisely but falls between 385 and 370.

—with which many of the most brilliant scientists of the fourth century were at one time or another associated, even though Plato's own primary aim was to train political philosophers—but also and more especially for the views he expressed on the basis and aims of scientific investigation.

Plato's relevance to our study lies, then, less in the particular scientific theories that he put forward than in what we may call his philosophy of science, and it is with this that we shall be chiefly concerned in this chapter. The view often taken concerning this aspect of his thought is, however, an extreme one. He has frequently been represented as an arch enemy of science. The philosophy based on the theory of Forms was—so it is argued—utterly antagonistic to science and constituted an important obstacle to its development, and passages in the *Republic* and the *Timaeus* in particular are often cited to show how hostile he was to specific scientific disciplines. To find out how much truth there is in this view we may turn first to the interpretation of some controversial texts in the *Republic*.

In book VII Socrates is describing the training of the philosopher-kings who are to be the guardians of the ideal state, and he considers in turn the role of arithmetic, plane and solid geometry, astronomy and acoustics in higher education. His remarks on astronomy are particularly provocative. When he first suggests that astronomy should be one of their propaedeutic studies, Glaucon is made to misunderstand him in two ways. First, Glaucon assumes that astronomy has been recommended because it is useful.

> Skill in perceiving the seasons, months and years is useful not only to agriculture and navigation, but also just as much to the military art (527d).

But to this Socrates remarks:

> I am amused that you seem to be afraid lest the many suppose you to be recommending useless studies.

But Glaucon's second attempt to justify astronomy is no better than the first:

> In place of the banal recommendation of astronomy, for which you reproved me just now, I shall now recommend it after your manner. For I think it is clear to everybody that this study at least compels the soul to look upwards and leads it away from things here to things there (528e f).

Once again Socrates shows Glaucon how mistaken he is. The true value of astronomy lies in its ability to direct the attention of the soul not to any visible objects, but to certain invisible realities. He describes the stars themselves as 'decorations', *poikilmata*. They are admittedly the most beautiful and exact of visible things, but they fall far short of the truth which is to be apprehended by reason and thought alone. We must use the stars as 'patterns' to help us in our study, just as we might use geometrical diagrams in doing geometry. But

> anyone experienced in geometry who saw such diagrams would grant that they are most beautifully constructed, but think it absurd to examine them seriously as if one could find in them the truth concerning equals or doubles or any other ratio (529e f).

A true astronomer will not imagine that the proportions of night to day or of both to the month, for example, remain unchanging in objects that are visible and material. Rather:

> it is by using problems . . . , as in geometry, that we shall study astronomy too, and we shall let the things in the heavens alone, if we are to participate in the true astronomy and so convert the natural intelligence in the soul from uselessness to a useful possession (530bc).

To interpret this passage correctly it is essential to examine its context. The over-all purpose of the guardians' training is to lead them away from the visible to the intelligible world, to make their souls cultivate reason rather than sensation. Throughout book VII the criterion used to decide whether a study is suitable for educating the guardians is: does it encourage abstract thought? In this context Plato naturally emphasises the distinction between an observational, and an abstract, mathematical astronomy, and his comments on the correct way of doing astronomy should be judged in the light of the general educational purposes which he suggests it can serve. Moreover the distinction he draws between a purely observational astronomy and a study that proceeds by means of problems is a useful and important one, and indeed the modern scientist would agree that his primary concern is the latter. Finally Plato is perfectly correct both when he suggests that the heavenly bodies do not in fact conform exactly to mathematically determined courses, and when he implies that the heavens are not completely unchanging.

There is, then, a good deal that can be said in defence of Plato's position in this passage in the *Republic*. At the same time, however, many of his remarks are exaggerated, and several are not only vague but also dangerously ambiguous. Take the passage where he compares the heavens with geometrical diagrams and remarks that it is absurd 'to examine them seriously as if one could find in them the truth concerning equals. . . .' Here he may be making the simple and perfectly unobjectionable point that diagrams are necessarily imprecise: one does not get out a ruler to determine the length of the hypotenuse of a right-angled triangle whose shorter sides are three inches and four inches. But he might also be taken to be advancing a much more extreme thesis, that to examine the diagrams at all is useless. In this and other passages there appears to be an assimilation or confusion of two ideas that should have been kept distinct—the true and obvious point that we cannot observe the mathematically determinable courses of the heavenly bodies as such, and the controversial thesis that it serves no use at all to observe the heavenly bodies. Plato evidently recognised that his ideal astronomy represented a radical departure from the usual way of studying astronomy in his day. But in advocating his new astronomy he speaks as if he thought it necessary not merely to distinguish it from observational astronomy, but to run the latter down. Similar points can be made concerning his discussion of acoustics, too, where he again advocates a mathematisation of the science, but again argues, with even less justification than in astronomy, against observational methods, which he describes at one point (531a) as useless labour.

Apart from the *Republic* the main work we must consider to assess Plato's place in the development of Greek science is the *Timaeus*. This contains a detailed cosmology which is described as an *eikos mythos* or an *eikos logos*, and our first problem is what is meant by this. To consider the account as what we should call a myth would be wrong. It is true that many of the details, particularly those relating to the work of the Craftsman (*demiourgos*), are figurative and should not be interpreted literally. Nor should we assume that the order in which the creation of different things is described is meant to correspond to a historical sequence of events. Nevertheless the cosmology is a probable account or story, rather than a myth or fiction.

Plato leaves us in little doubt about his intentions. At 27d *ff* the chief speaker in the dialogue, Timaeus— who may be taken to be, in the main, a spokesman for Plato himself—explains what sort of account he will give. He first distinguishes between the eternally existing Forms and the changing world of becoming: the former are the models from which the latter is copied. And he next distinguishes the types of account that are appropriate to each. He demands that statements about the unchanging reality, the Forms, should be irrefutable, at least as far as possible. Then about the changing world of becoming he says:

> If in our discussion of many matters . . . we are not able to give perfectly exact and self-consistent accounts, do not be surprised: rather we should be content if we provide accounts that are second to none in probability (29c).

The cosmology in the *Timaeus* is not an exact account. Indeed Plato believes that the nature of the subject-matter precludes this. On the other hand he does claim that it is the *best account possible*, given that it deals with the world of becoming.

Plato repeatedly points out that no exact account of the world of becoming is possible and says that no such account can be asserted to be true, and in this he is, one may say, much less dogmatic than most earlier, and indeed than most later, Greek cosmologists. But the reason for this is not that he believes that judgement should be withheld until more evidence has been assembled, but rather that he holds on principle that no account of the world of becoming can, under any circumstances, be certain. The effect of his undogmatic cosmology was undoubtedly not to encourage more empirical research, but rather the reverse. Just as the *Republic* depreciates the role of observation in the propaedeutic studies of the guardians, so too the *Timaeus* shows a similar impatience with those who hoped to resolve the problems of physics by empirical methods.

But if the philosopher's primary concern is the world of Forms, what role does the study of the world of becoming have? One passage in the *Timaeus* suggests that Plato considered this study rather in the nature of a recreation. When we turn aside from discussing eternal things, we can gain pleasure from probable accounts of coming-to-be, for this is a 'moderate and intelligent pastime' (59cd). Yet if this is a 'pastime', it is not one undertaken frivolously. At 68e *ff* we are given good reasons for studying both what are called the 'divine', and the 'necessary', causes. They should seek

the divine in all things for the sake of gaining a life that is as happy as our nature permits, but also the necessary causes, for the sake of the divine, recognising that without them it is impossible to apprehend by themselves alone the divine things that are the objects of our study.

The ultimate justification for the study of the world

of nature is an ethical one, but we can be more precise than this. If we ask why Plato embarked on a detailed account of the world of becoming, the nature of the cosmology he produced provides the chief clue to the answer. Throughout his account he stresses the role of an intelligent, purposive agency in the universe. This all-pervasive teleology—the belief in the element of design in nature—is often mentioned as one of the un-scientific or anti-scientific features of Plato's physics, and it is true that many of his ideas about the functions served by various parts of the body, for example, strike us as quaint. Thus the lungs are described at one point (70c) as a buffer to provide relief for the throbbing of the heart, and in another passage (72c) the spleen, whose function is said to be to keep the liver free from impurities, is compared with a napkin set beside a mirror to keep it clean. Yet paradoxically it is because of the teleological bias in his account that we can be sure that his cosmology is intended seriously. Plato's main motive for doing what we should call natural science is to reveal the operations of reason in the universe. While he often contrasts the world of becoming unfavourably with the Forms, he nevertheless repeatedly asserts that this is the best possible created world. It is the fairest of things that come to be (29a), its maker is good (30ab), it is fashioned after the most perfect model and it is as like that model as possible (30d, 39e). It is, as the last sentence in the *Timaeus* puts it, 'a perceptible god, greatest and best and fairest and most perfect'.

The three main elements in Plato's cosmological scheme are, then, first the Forms, second the particulars modelled on them, and third the agent who does the modelling, the Craftsman. The Craftsman does not create the world in the sense of creating the matter of which it is made. Rather he is imagined as taking over pre-existing matter and imposing order on its disorderly movements. He is not omnipotent, but achieves the best results possible. Plato describes on the one hand the works of Reason, and on the other what comes to be 'through necessity'—the effects of what is called the 'wandering cause'. This is not an active principle opposing the Craftsman as a force of evil, but rather the passive resistance that disorderly matter offers to the Craftsman's design. The role of necessity is best illustrated with an example, the creation of the head. Dense bone and flesh, we are told (74e f), make for insensitivity. So to procure for man a noble and intelligent life, the craftsman-gods choose to leave his brain covered with only a thin layer of bone, even though this means that man's life is shorter that it would have been if the head had been protected by a thicker covering of flesh and bone. Evidently long life and keen intelligence cannot *both* be achieved: so at this point reason sacrifices the lower for the higher end. The example is bizarre, but it illustrates how the results that reason achieves are not the best absolutely, but the best given the limitations imposed by the nature of the material with which it has to work.

The cosmology of the *Timaeus* is set in a complex and distinctive philosophical framework. Plato's reputation ensured that the *Timaeus* became an influential document, but we must now try to evaluate the specific theories and explanations it advances. The first problem that we must consider here is how far, indeed, they are Plato's own theories. It is clear that he took over a great deal from his predecessors and contemporaries even though he refers to none of his sources by name. We can identify specific debts to Empedocles, the Pythagoreans and the atomists, and the biological sections owe much to the Hippocratic writers and to such men as Alcmaeon, Diogenes of Apollonia and Philistion of Locri. Yet it would be a mistake to conclude, as some commentators have done, that the natural science of the *Timaeus* is simply a collection of other people's ideas. Any scientist must build to some extent on earlier work, and although Plato's debt to earlier theorists is particularly great, he does not merely copy or repeat their doctrines, but modifies and adapts them, and on occasions he introduces important ideas that appear to be original.

His most notable contribution to the physics of his day lies in his doctrine of the ultimate constituents of matter. At 49a *ff* Timaeus analyses the conditions of change. He draws attention to the instability of what we perceive and says that we should not describe fire, water and so on as if they were definite, stable objects. He distinguishes *what* comes to be from that *in which* it comes to be, calling the latter the 'receptacle' of becoming. This was an important idea which influenced Aristotle's conception of the substratum and made an original contribution to the philosophical analysis of change. And then Timaeus proceeds to give an account of what comes to be. The theory he presents borrows much from both Empedocles and the atomists. But in combining ideas from these and other sources Plato produced what is, in certain respects, a new solution to the problem of the constitution of matter.

Like Empedocles, he considered every natural substance to be a compound of the four simple bodies, fire, air, water and earth. But, unlike Empedocles, he did not stop at that point in his analysis, but identified each of the four simple bodies with one of the regular solids, fire with the tetrahedron (4 faces), air with the octahedron (8), water with the icosahedron (20) and earth with the cube (6)—the fifth regular solid, the dodecahedron, is mentioned, but not identified with a simple body, at 55c. Although we cannot be sure at what stage the five regular polyhedra were discovered by Greek mathematicians, we may assume that their construction and properties continued to exercise Plato's contemporaries, and Plato himself is no doubt drawing on their work. He constructs the four figures he uses for the simple bodies from two kinds of basic triangles, the right-angled isosceles and the half equilateral. Thus right-angled isosceles triangles can be combined in various ways to form a square—the face of the cube. And similarly half-equilaterals combine to form the equilateral triangles that are the faces of the tetrahedron, octahedron and icosahedron. Interestingly enough, Plato did not choose the simplest construction for his figures. Instead of making the square with two right-angled isosceles triangles (A in diagram 2), he used four (B), and instead of using two half-equilaterals to make the side of the other solids (C), he used six (D). The reason for this is

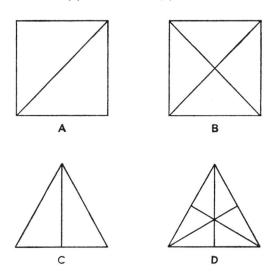

Diagram 2   The geometry of Plato's elements.

obscure: it may be connected with the construction used to define the centre of the solids in question, or alternatively with Plato's belief that each simple body exists in different forms according to the size or grade of the units that compose it. He may have wished the complication in the construction to suggest that the basic triangles of each body can be assembled in different ways to produce the different 'isotopes' of that body, as for example right-angled isosceles triangles being put together not only in fours, but also in twos and other powers of two, to produce cubes of varying size corresponding to different forms of the simple body, earth (see diagram 3).

How does Plato's theory compare with its main rivals,

those of Empedocles and the atomists? First, it is more economical than Empedocles', which required four distinct types of matter. And in recognising that changes take place between fire, air and water, at least, it evades some of the empirical objections to which Empedocles' theory was open. As we saw, Empedocles did not allow the transformation of one root into another, and yet it is a matter of common experience that water, for

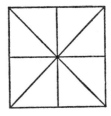

Diagram 3   To illustrate the arrangement of two, four and eight right-angled isosceles triangles to form squares of three sizes corresponding to three 'grades' of earth.

instance, on being heated to boiling point, becomes vapour, that is, in the Greek view, 'air', and that this air may recondense and become water again.

The differences between Plato's theory and that of the atomists are also instructive. He owed to them the idea that the varieties of sensible objects can be reduced to differences in the shapes and sizes of particles that are themselves homogeneous in substance. But while they thought of the basic particles of matter as solid, Plato suggested that the primary solids are in turn composed of plane surfaces made up of his two basic kinds of triangles. Secondly, while the atomists postulated the existence of a void, Plato denies this and evidently saw that in a plenum motion is possible provided it is (*i*) instantaneous and (*ii*) cyclical. According to his doctrine of the 'circular thrust', the movement of A pushes B which pushes C and so on, and Z the last term in the series itself

76

pushes A. Third, and most important, whereas the atomists apparently postulated an indefinite variety of atomic shapes and sizes and described the interactions between them in only the most general terms, Plato attempted a definite and specific account both of the shapes of the primary bodies and of the transformations that take place between them. At 56d ff he makes a number of specific suggestions about, for example, how water may be decomposed into fire and air. Thus the icosahedron of water may become two octahedra of air and one tetrahedron of fire, the original solid of twenty faces being broken down into two solids of eight faces each and one of four.

Many details of Plato's theory remain obscure. How can the primary bodies disintegrate and recombine to form other figures? How indeed can the primary solids be described as bodies at all, when they are geometrical entities constructed from plane surfaces? There are, too, many fanciful and arbitrary elements in Plato's doctrine. Thus having assigned four of the five regular solids to the four simple bodies, and not finding any other task for the fifth, the dodecahedron, he identified it with the twelve signs of the zodiac. Again when we find that earth is excluded from the transformations that affect the other primary solids, the reason for this does not lie in any real or supposed empirical data, but it is a direct consequence of the geometry of the theory, once earth had been identified with the cube. Yet what Plato did, that the atomists themselves never attempted, was to propose a precise geometrical account of the shapes of the primary bodies and to reduce the changes that take place between them to mathematical formulae. Many of his ideas, like those of Leucippus and Democritus, are still based on crude physical analogies—as for example the identification of fire with the tetrahedron and earth with the cube—but he attempted to carry the geometrisation of atomism much further than the original atomists themselves did.

The *Timaeus* contains a wealth of detailed physical

and biological doctrines, including, for instance, an ingenious account of respiration and a long discussion of the causes of diseases, and much as these theories owe to earlier writers, they are far from being entirely unoriginal. Yet, as we have noted, in the long run it was not the specific theories and explanations in the *Timaeus* that were to prove most influential, so much as the philosophical ideas that characterise Plato's whole approach to the inquiry concerning nature.

The two main doctrines that mark out Plato's view are his teleology and his relative evaluation of reason and sensation, and each of these doctrines has both a positive and a negative aspect. First, one of the unfortunate consequences of his teleology—we might say—is that the amount of attention he devotes to different problems reflects his notion of the extent to which the phenomena in question manifest order and rationality. Thus while quite a long discussion is devoted to astronomy (see pp. 83 *ff*), the *Timaeus* shows only a cursory interest in what we should describe as the problems of mechanics. Again while in his sections on human anatomy some elaborate suggestions are put forward concerning the functions served by the various organs in the body, zoology and botany are almost entirely ignored: there is scarcely a mention of the other species of animals besides man until the very end of the dialogue, where he refers very briefly to their different kinds principally in order to suggest that they originated from degenerate human beings.

Yet Plato's teleology provides his main motive for doing both cosmology and natural science. It is because natural phenomena show evidence of order that they are worth studying. Moreover, while the study of design in nature is Plato's chief concern, he says that they must seek not only the 'divine' causes, but the 'necessary' ones as well, the latter for the sake of the former. So it is that he undertakes a far more detailed and elaborate account of natural phenomena than we might have expected in view of his estimate of the relative importance of the two worlds, of being and of becoming. This was not the first time in the history of Greek science, and it was far from being the last, that an inquiry originally undertaken for what were, broadly speaking, ethical motives led not merely to the proposal of morally and aesthetically satisfying cosmological images, but also to certain developments in physical and biological theories.

Secondly, his preference for reason over sensation and observation may also be said to have had both beneficial and unfortunate results from the point of view of the inquiry concerning nature. His methods of investigating natural phenomena may, in certain respects, be contrasted unfavourably with those of some of his more empirical contemporaries, particularly among the medical writers. It was not Plato's way to undertake detailed empirical research in connection with his accounts of causes, and sometimes—for example in anatomy—much might have been gained had he done so. Although his more provocative remarks denigrating the use of the senses should be interpreted as suggesting merely that observation is inferior to abstract thought, and not as suggesting that observation is completely worthless, their effect, in some quarters at any rate, was still to discourage empirical investigations.

Yet here too the positive aspects of Plato's position should not be ignored. He is right to insist that the scientist's inquiry is directed to the discovery of the abstract laws that lie behind the empirical data. His belief in the mathematical structure of the universe— taken over and developed from the Pythagoreans—and his conception of an ideal, mathematical astronomy and physics are his two most important and fruitful ideas, and the fact that we nowadays take both of them so much for granted does not diminish, but rather heightens, Plato's achievement in being their most powerful exponent in antiquity.

7

Fourth-Century Astronomy

THE greatest achievements of early Greek science lie in astronomy. This was the one science to which mathematical methods were applied, and applied with a good deal of success, before the end of the fourth century. To appreciate the nature of this advance, we must first recapitulate briefly some features of the earlier history of astronomy.

Greek attempts to construct a mechanical model for the heavenly bodies go back, as we noted in Chapter 2, to Anaximander. Yet the fact that he represented the stars as below the sun and moon indicates the primitive nature of his system, and when he suggested that the rings of the sun, moon and stars are spaced at regular intervals—their diameters being twenty-seven, eighteen and nine earth-diameters respectively—this was not so much the outcome of an attempt to 'save the phenomena' by reducing them to mathematical laws, as a reflection of Anaximander's liking for symmetry and of his belief in the special importance of the number three. The same considerations also underlie his suggestion that the earth is three times as broad as it is deep, which was, even more clearly, a theory that could have had no basis in observation.

Several of the more obvious facts that Anaximander's system ignored, such as the distinction between fixed stars and planets, were pointed out before long. At the same time some of his errors were remarkably persistent, including both the doctrine of a flat earth and the idea that the sun is the most distant of the heavenly bodies, both of which recur, for example, in Leucippus. The task of assessing the Presocratics as astronomers is especially difficult as our sources are particularly unreliable in reporting theories in this area. The doxographers

were fond of attributing specific astronomical discoveries, such as the obliquity of the ecliptic,[1] or the
identity of the Morning and the Evening Star—that is
the planet Venus—to individual Greek theorists, although they frequently disagree on who the discoverer
was. Moreover in many cases the Babylonians had long
been familiar with the data in question, and although
we know little about the transmission of astronomical
information to Greece, it is often just as likely that
the Presocratic philosophers derived their knowledge
directly or indirectly from the East as that they made the
discoveries independently.

This is not to say that the Greeks failed to conduct
observations of the heavenly bodies for themselves.
Apart from any theoretical motives, there were at least
two practical reasons why frequent, even if only rough
and ready, observations should be carried out. First, in
Greece as elsewhere, the farmer's year was governed by
observations of the rising and setting of certain constellations, as we see from Hesiod, for example:

> But when Orion and Sirius come into mid-heaven, and
> rosy-fingered Dawn looks on Arcturus [that is when
> Arcturus rises just before the sun], then, Perses, pick all
> your grapes and carry them home (*Works* 609 *ff*).

And another, more important, stimulus to the study of
celestial phenomena was provided by the need to regulate the calendar. As already noted (p. 6), considerable
progress was made, during the fifth century, in determining the relation between the lunar month and the
solar year. By 432 Meton of Athens had made a tolerably
accurate calculation of the right number of intercalary

---

[1] The ecliptic, called by the Greeks the 'oblique circle' or the
'circle through the zodiac', is the great circle of the celestial sphere
which is the apparent orbit of the sun: it is inclined to the celestial
equator at an angle of about $23\frac{1}{2}°$ (see diagram 4 on p. 86) and it
gets its name from the fact that eclipses can take place only when
the moon is at or near this line.

months in a nineteen-year cycle, even though his fellow-citizens did not make use of his results to reform their own civil calendar. And by about the same time another important astronomical datum was also definitely recognised, namely that the four seasons, measured by the solstices and the equinoxes, are of unequal length: we find definite estimates of the lengths of the seasons ascribed to Meton's contemporary Euctemon in the second-century B.C. astronomical papyrus known as the *Ars Eudoxi.*

The fifth century also saw a proliferation of theories concerning the relations between the main heavenly bodies, the most notable of which is the Philolaic system (outlined in Chapter 3) in which the earth is removed from the centre of the universe. Yet although this theory constituted a remarkable advance on earlier ideas, particularly in that it identified the five principal planets[1] and assigned them to separate circles underneath the sphere of the fixed stars, there is no evidence that either it or any other late fifth- or early fourth-century system attempted a precise account of the movements of the various heavenly bodies. It is this that distinguishes the later fourth-century theories and marks an epoch in the development of astronomy.

The credit for the first such mathematical account of the movements of the heavenly bodies belongs to Eudoxus of Cnidus, a younger contemporary of Plato and an associate of his in the Academy. But Plato's own part in this development should also be mentioned. Several dialogues, especially the *Republic* and *Timaeus,* contain passages that deal with astronomical questions, although the language in which they do so is often both high-flown and obscure. In the myth of Er there is a description of the Spindle of Necessity, the whorl of which is said to consist of eight separate whorls fitting closely together—like a nest of Chinese boxes (*Republic* 616c ff). The outermost of these whorls represents the

[1] The five planets known to the ancients were Saturn, Jupiter, Mars, Venus and Mercury.

sphere of the fixed stars, and the other seven the circles of the planets, sun and moon. The Spindle turns round as a whole with one motion, and 'within the whole as it revolves the seven inner circles revolve slowly in the opposite sense to the whole' (617a). The speeds with which the circles move vary: thus the eighth circle, that of the moon, moves fastest, and the seventh, sixth and fifth—those of the sun, Venus and Mercury—move with the same speed. The breadths of the whorls also differ—these presumably correspond to the distances between the circles of the lower heavenly bodies—and the text also remarks on the different quality of the light of the different circles. In the *Timaeus* (36c *ff*), two main types of movement are again distinguished, 'motion along the circle of the Same', and 'motion along the circle of the Other', and reference is now made to a point that was ignored in the *Republic*, namely the obliquity of the circle of the Other, that is, the ecliptic. Finally *Timaeus* 40bc should also be mentioned, as this passage has been taken to imply that Plato held that the earth rotates on its axis. Here the earth is described as 'winding round' (*illomenen*) the pole: the expression is obscure, but on the most likely interpretation all that it implies is that the earth is endowed with a force to counteract the movement of the circle of the Same—so that in relation to absolute space the earth stands still. At least there is no doubt that the diurnal revolution is explained as the effect of the circle of the Same, and not as the result of any movement of the earth.

These passages in Plato are the earliest extensive Greek astronomical texts to have survived, and although the problems of interpretation are severe, they contain the first clear extant statements of at least two important doctrines. First, he distinguished between two kinds of movement, (*i*) the movement of the sphere of the fixed stars, which is shared by all the heavenly bodies, and (*ii*) the independent movements of the sun, moon and planets along the oblique circle of the ecliptic in an opposite sense to that of movement (i). Secondly, he recognised that Venus and Mercury move with the same mean speed—that is, as we should say, angular velocity —as the sun: the two lowest planets are never far from the sun, and all three complete a revolution of the zodiac in (approximately) one year.

But even more influential than Plato's own forays into astronomical theory were the ideas he expressed on the type of account that astronomers should aim at. We have seen that in the *Republic* (528e *ff*) he recommends an ideal or mathematical astronomy in place of an observational one, and he is evidently aware that what he is advocating represents a radical departure from the usual way of doing astronomy: 'It is by using problems . . . , as in geometry, that we shall study astronomy too', even if this method is 'many times more laborious than the present mode of doing astronomy' (530bc).

Then, apart from the general injunction to study astronomy 'by using problems', Plato is reputed to have been responsible for formulating the particular problem that was to remain the chief preoccupation of astronomers for many centuries, namely, that of planetary motion. This is according to a late, but not totally implausible, report that Simplicius attributes to a writer named Sosigenes (second century A.D.), which has it that Plato put this question to students of astronomy: 'By the assumption of what uniform and orderly motions can the apparent motions of the planets be accounted for?' To be sure, the search for regularity and order is, in a sense, as old as theoretical astronomy itself. But the question as reported is not merely a general recommendation to study the planets. First, it is recognised that the apparent movements of the planets present anomalies that require explanation. To an observer on earth they 'wander', as indeed is implied by their Greek name *planetes*, which is derived from the verb *planaomai*, wander, and simply means 'the wanderers'. Secondly, it is assumed that the irregularities in their movements are only apparent and are produced by a combination of movements that are themselves uniform and orderly. And thirdly—although

84

this is not explicit in the text quoted by Simplicius—the motion that is supremely 'uniform and orderly' is circular motion. The problem could be rephrased, then, as being how to combine various uniform, circular motions in such a way that their resultant corresponds to the observed movements of the planets. Although the conditions of the problem changed, particularly when the requirement of circular motion was abandoned by Kepler, the explanation of the observed movements of the planets was to remain the chief issue in astronomy right down to Newton.

Plato's main contribution was to insist that astronomy is an exact, mathematical science, and the later fourth-century astronomers took up his challenge and attempted comprehensive accounts of all the main features of the movements of the heavenly bodies. From the point of view of an observer on earth, the main phenomena may be summarised under five heads. First, all the heavenly bodies circle the earth from east to west once in about twenty-four hours—the effect, we should say, of the earth's daily rotation about its axis. Secondly, different constellations are visible from a given point at different seasons in the year, but the same constellations appear in approximately the same places at the same seasons in successive years. Thirdly, the sun's position in relation to the stars (determined by observing the constellations that set just after, or rise just before, it) changes regularly. It moves from west to east through a band of constellations—the zodiac—completing a single revolution, that is coming back to the same constellation, in about a year—the effect, as we should describe it, of the earth's circling the sun. Fourthly, the moon and each of the planets also move from west to east through the constellations, indeed through the same constellations as the sun: the paths of the moon and the visible planets do not deviate by more than eight degrees from the course of the sun itself, the great circle known as the ecliptic (see diagram 4). However, the period they take to complete a single revolution of the zodiac varies:

85

Saturn takes nearly thirty years, Venus and Mercury about one year, while the moon makes a complete revolution in about a month.

Finally, when a planet's position is tracked over a number of months, the irregularities known as the stations and retrogradations soon become apparent. Its

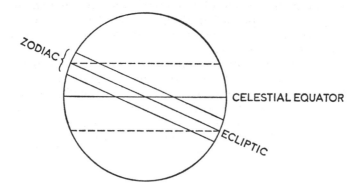

Diagram 4   The ecliptic.

easterly movement through the constellations is occasionally interrupted. For a number of days the planet's position in relation to the stars hardly alters. It then begins to move back through the constellations from east to west for a time. Once again it seems to stand still in relation to the stars, and finally it resumes its usual easterly course (see diagram 5). It was this feature of the movements of the planets that constituted the chief difficulty that the later fourth-century astronomers set themselves to explain.

Eudoxus' solution, preserved for us by Aristotle and Simplicius although his own work is lost, was most ingenious. He suggested that the complex apparent paths of the sun, moon and planets were, in each case, produced by the simple circular movements of a certain number of concentric spheres. The earth is at rest at the
86

common centre of all the spheres, but their axes are inclined to one another and they rotate at different, though uniform, speeds.

Thus for each of the five known planets he postulated four such spheres, the planet itself being imagined as

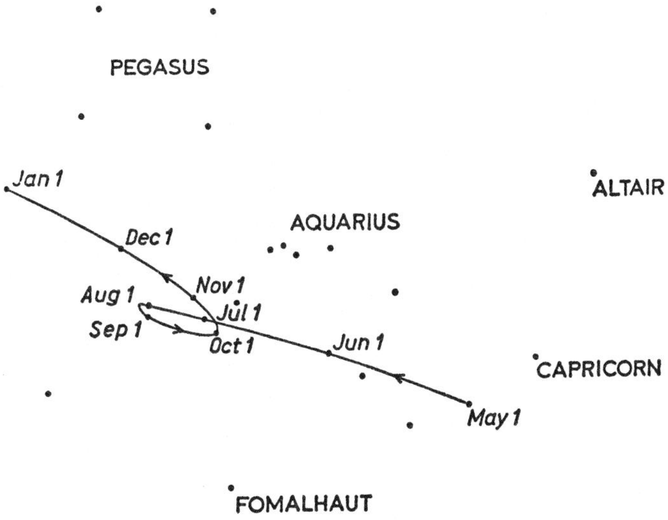

Diagram 5   Path of Mars from 1 May 1956 to 1 January 1957. The planet was stationary on 11 August, and again on 12 October: between those two dates its motion was retrograde.

lying on the equator of the lowest or innermost sphere (see diagram 6). The first or outermost sphere (1), in each case, 'moves', as our sources put it, 'with the movement of the fixed stars'—that is, it accounted for the phenomena caused by the daily rotation of the earth. The poles of this sphere lie on a north–south axis and it rotates from east to west, completing one revolution in every twenty-four hours.

The second sphere (2), in each case, produced the apparent movement of the planet along the zodiac. The axis of this sphere is perpendicular to the plane of the

87

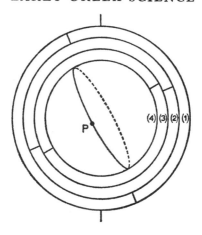

Diagram 6   Eudoxus' theory of concentric spheres. The planet (P) is on the equator of sphere (4), out of the plane of the rest of the diagram.

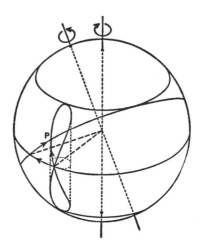

Diagram 7   To illustrate the 'hippopede' of Eudoxus. From Neugebauer, *Scripta Mathematica*, no. 19 (1953), p. 229.

ecliptic. It rotates from west to east, but the speed with which it does so differs from planet to planet. Eudoxus' estimates of the periods of zodiacal revolution of each of the five planets are given in round figures by Simplicius, and they correspond closely to the modern values quoted by Heath.[1]

But the most remarkable part of the theory is the movement of the two lowest spheres (3 and 4), which were used to explain the planet's stations and retrogradations. The poles of the third sphere lie on the circle of the ecliptic, while the axis of the fourth sphere is inclined to that of the third at an angle that varies from planet to planet. These two spheres rotate at equal speeds but in opposite directions, and their combined movement produces the curve that Eudoxus named the 'hippopede' or horse-fetter. This is an 8-shaped curve, which can be described as the intersection of a sphere with a cylinder touching it internally at a double point (see diagram 7), and when this closed curve is in turn combined with the movement of the second sphere carrying the planet along the zodiacal band it gives a fair approximation to the looping motions described by the planets, including the variations in their apparent speed as they approach and draw away from the stationary points.

For the sun and moon Eudoxus' theory was similar but rather less elaborate, since they do not exhibit the phenomena of stations and retrogradations. In each case he postulated three spheres. The first two of these correspond to the first two spheres of each of the planets, the outermost moving with the movement of the fixed stars, and the second moving about an axis perpendicular to the plane of the ecliptic. The third sphere was used to explain deviations from the ecliptic, which are indeed a notable feature of the moon's movement. The sun, however, suffers no such deviation: the idea that it did was an error into which Eudoxus was led by observing that the sun does not always rise at precisely the same

[1] *Aristarchus of Samos*, Oxford, Clarendon Press, 1913, p. 208.

point on the horizon at the summer and winter solstices.

The system as a whole shows great mathematical skill
and it managed to account fairly successfully for a wide
variety of phenomena without breaking Plato's rule that
only simple circular movements should be postulated.
Given the geometrical methods available at the time, it
was a remarkable feat to devise a combination of spheres
to produce a curve corresponding to the looping movement of the planets. Moreover, unlike earlier astronomers, Eudoxus was not content to leave his model at the
point where he had suggested in vague, general terms
how various phenomena might be explained, for he
evidently worked out his theory for each planet in some
detail. Although not all his figures are recorded, in most
cases he seems to have given definite estimates both for
the periods of revolution of the various spheres and for
the angles of inclination of their axes to one another.

Nevertheless at certain points, some of them important ones, the theory failed to account for the facts. Four
of the main difficulties may be noted briefly. First, each
hippopede always produced exactly the same curve: but
the observed retrogradations of each of the planets vary
in shape, size and duration (see diagram 8). Secondly,
while Eudoxus' figures provide the basis for a tolerable
solution to the observed retrogradations of Saturn and
Jupiter, they fail to do so for Mars and Venus. The
figure quoted in our sources for Mars' synodic period
(the period between two successive conjunctions with
the sun) is wide of the mark. But if the true figure is
taken, then so long as the lowest two spheres rotate in
opposite directions, no retrogradation can occur. And
although retrogradations will take place if Eudoxus'
own grossly inaccurate figure for the synodic period is
assumed, these still fail to tally with the observed
courses. Thirdly, the system failed to account for the
inequality of the seasons, even though—as already noted
(page 82)—this was certainly known to Euctemon some
time before Eudoxus. And fourthly, it failed to explain

variations in the apparent diameter of the moon and in the brightness of the planets—phenomena which later Greek astronomers correctly understood as being due to the fact that their distances vary. The theory of concentric spheres could not allow for differences in the distances of the sun, moon and planets from the earth, and indeed it was on this point, especially, that this type

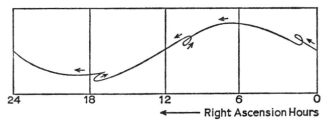

Diagram 8   Path of the planet Mercury in 1958, showing different shapes of retrograde loop. From R. A. R. Tricker *The Paths of the Planets* (London, Mills and Boon, 1967)

of model was to founder, to be abandoned in favour of the theory of epicycles and eccentric circles.

Such, however, was the prestige of Eudoxus' solution to the problem of planetary motion that rather than abandon his model entirely, his immediate successors tried to modify it in order to allow for some of the phenomena that it failed to explain. The first such modification was the work of Callippus of Cyzicus, a younger contemporary of Eudoxus, whose theory is again reported by Aristotle and commented on by Simplicius. Where Eudoxus had postulated a total of twenty-seven spheres—four for each of the five planets, three each for the sun and moon and one for the sphere of the fixed stars—Callippus increased this number by seven. The number of spheres for Saturn and Jupiter he left unchanged, these being, as I have already noted, the two cases where Eudoxus' model was most successful in describing the observed retrogradations. But he added one sphere to each of the other three planets and two each to the sun and moon.

91

Simplicius tells us that the two extra spheres postulated for the sun were designed to explain the third of the four major difficulties we noted in Eudoxus' theory, namely the inequality of the seasons, and this seems very likely. The *Ars Eudoxi* attributes to Callippus close estimates of the lengths of the four seasons as measured by the solstices and equinoxes. Beginning with the vernal equinox, he made the lengths of the four seasons 94, 92, 89 and 90 days respectively—figures which not only are more accurate than those ascribed to Euctemon, but have been calculated by modern experts to be correct to the nearest whole number for the period at which Callippus was making his observations (about 330 B.C.). The two extra spheres for the moon were probably similarly applied to explain inequalities in its motion along the zodiac, and the extra sphere postulated for each of the lower three planets was presumably introduced to get better approximations to their observed retrogradations.

The theories of Eudoxus and Callippus were purely mathematical constructions. Neither astronomer said anything about the mechanics of the heavenly movements, about the nature of the concentric spheres or about how movement was transmitted from one to the other. The next modification of Eudoxus' model was the work of Aristotle who attempted to use this mathematical theory as the basis of a mechanical system. He was interested not only in trying to reduce the observed paths of the planets to combinations of simple circular movements, but also and more especially in the problem of the transmission of movement from the outermost sphere of heaven to the region below the moon.

For movement to take place, the spheres must, Aristotle believed, be in contact with one another. But as soon as Eudoxus' spheres are conceived as connected in a mechanical system, the movement of each of the heavenly bodies will be affected not only by its own spheres, but by all the spheres above it. So Aristotle was forced to introduce a number of 'reacting' spheres, the purpose of which was to cancel out the movements of certain of the primary spheres: each reacting sphere had the same axis as the primary sphere it counteracted and rotated at the same speed in the opposite sense. He thought that in every case except the moon, which being the lowest of the heavenly bodies needed no spheres to counteract its movement, the reacting spheres would be one fewer than the primary spheres which he took over from Eudoxus and Callippus. Using the Callippan version of the system, he made the total number of spheres needed fifty-five, or fifty-six including the sphere of the fixed stars itself. But he also considered the possibility that the extra spheres that Callippus had postulated for the sun and moon were not necessary, in which case the total number of spheres would be rather less. Our text gives the number as forty-seven, but this seems to be a mistake. If we deduct two spheres each for the sun and moon, and the corresponding two reacting spheres for the sun (remembering that the moon has no such reacting spheres) the figure will be six, not eight, less than the original fifty-five, that is forty-nine: so we must assume either that the text is corrupt or that Aristotle himself slipped up, perhaps forgetting that he postulated no reacting spheres for the moon.

This is far from being the only unsatisfactory feature of Aristotle's astronomical digression in the *Metaphysics* (1073b 10 *ff*). Thus the number of spheres postulated seems, in any event, too high, since on analysis it turns out that the first primary sphere of each planet exactly reduplicates the motion of the last reacting sphere of the planet above: both move with the motion of the fixed stars.

Nevertheless Aristotle makes it clear that on astronomical matters he speaks as a layman. When he embarks on his discussion of the number of spheres, he says he will report what certain of the astronomers say, so that our thought may have a definite number to grasp:

> But, for the rest, we must partly investigate for ourselves, partly learn from other investigators, and if those

who study this subject form an opinion contrary to what we have now stated, we must esteem both parties indeed, but follow the more accurate (*Metaphysics* 1073 b 13 *ff*).[1]

His account is, in fact, a provisional one and he denies any claim to have demonstrated it.

Despite the modifications introduced after Eudoxus, the model of concentric spheres failed to save all the phenomena: in particular no number of extra spheres could account for the noticeable variations in the brightness of the planets which suggested that their distances from the earth were not constant. It was superseded by the theory of epicycles and eccentric circles, but as this was a product of the third century it falls outside our period. One further fourth-century astronomer must, however, be mentioned. This is Heraclides Ponticus (so-called because he was a native of Heraclea on the Black Sea), who was a contemporary of Aristotle, and, like both him and Eudoxus, a pupil or associate of Plato. Historians of astronomy have seen him as the author of two important doctrines, the axial rotation of the earth, and the idea that Venus and Mercury move round the sun as centre, but while the former attribution may be accepted, the latter is open to question. Our sources for Heraclides' astronomy are late, scanty and confused. Moreover, it seems likely that he did not propose a definitive and comprehensive astronomical system, so much as suggest various often contrasting hypotheses as possible ways of accounting for the phenomena.

The doctrine of the axial rotation of the earth may owe something to earlier theorists. As we noted (p. 83), Plato used the term *illomenen*, 'winding round', of the earth in the *Timaeus*, but the correct interpretation of that passage seems to be that what he ascribed to the earth was merely a force to counteract the movement of the circle of the Same—so that in relation to absolute

---

[1] From the Oxford translation, *The Works of Aristotle translated into English*, edited by W. D. Ross (Oxford, Clarendon Press), *Metaphysics*, W. D. Ross (Vol. VIII, 2nd ed., 1928).

space the earth remains at rest. But whereas Plato still definitely assumed that the circle of the fixed stars moves, the doctrine ascribed to Heraclides by several of our sources goes much further. They suggest that he argued that the phenomena might be explained on the supposition that the heavens are at rest, while the earth rotates on its axis once in every twenty-four hours. Thus according to a text in Simplicius (*Commentary on Aristotle's On the Heavens*, 519, 9 *ff*), 'Heraclides supposed that the earth is in the centre and rotates while the heaven is at rest, and he thought by this supposition to save the phenomena,' and this report is borne out by other passages in Simplicius and other sources. The importance of this hypothesis is obvious: it implied a considerable saving in the number of celestial movements that had to be postulated. Yet like Aristarchus' heliocentric theory, it found little favour in antiquity, and for a similar reason: the ancient astronomers argued that if the earth underwent motion of any sort in space, this would have marked effects on the movements both of falling bodies and of the clouds, whereas no such effects are observed.

The second doctrine that is commonly attributed to Heraclides is that Venus and Mercury move round the sun as centre, and if indeed this was his view, it would provide a definite link between the theory of concentric spheres and the later model of epicycles and eccentrics, for according to this doctrine, while the sun still circles the earth, Venus and Mercury travel on what came to be known as epicycles round the sun (see diagram 9). The doctrine in question is certainly stated in a number of texts. Vitruvius (IX, 1, 6), for instance, says that 'The planets Venus and Mercury make their retrogradations and retardations about the rays of the sun, forming, with their orbits, a wreath about the sun itself as centre.' Yet neither Vitruvius, nor any other source that expresses the doctrine clearly, mentions its authorship or date, and the ascription of it to Heraclides rests on the testimony of a passage in Chalcidius' *Commentary on the Timaeus* which is particularly unsatisfactory: not only is the passage in question vague and unclear, but even the advocates of the view that it provides evidence that Heraclides maintained the circumsolar movement of Venus and Mercury must grant that Chalcidius has misrepresented that doctrine in certain vital respects.

But although the evidence that would enable us to

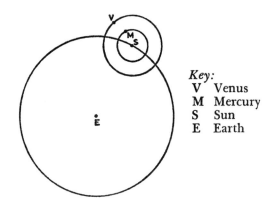

Key:
V   Venus
M   Mercury
S   Sun
E   Earth

Diagram 9   To illustrate the movement of Venus and Mercury on epicycles round the sun.

identify the author of this doctrine is not forthcoming, it is not impossible that it was put forward not long after Eudoxus. Plato had already suggested a connection between the movements of the two lowest planets and the sun, when he stated that all three 'move with the same speed'. Eudoxus too recognised that the sun, Venus and Mercury make a complete revolution of the constellations of the zodiac in a year, and we are explicitly told that he maintained that the poles of the third sphere are the same for both Venus and Mercury: on the most natural interpretation, the centre of the hippopede is the sun in the case of both these planets. It is clear that the geometry of the theory of epicycles and eccentric

96

circles was not worked out until Apollonius of Perga towards the end of the third century B.C. But it is possible that the theory of epicycles arose from the general application of a doctrine which had originally been suggested to explain the movements of the two lowest planets.

Both observational and theoretical astronomy made rapid progress in the fourth century. The increased accuracy of observations may be judged, for example, from the better estimates that were given of the lengths of the seasons. The first really comprehensive Greek star-catalogue was the work of the great second-century B.C. astronomer Hipparchus of Nicaea, but before that Eudoxus had attempted a description of the stars which formed the basis of the *Phaenomena,* the astronomical poem composed by Aratus of Soli in the early third century. Eudoxus' discussion of the motion of the planets presupposes extensive and careful observations of their positions in relation to the fixed stars. These observations had, furthermore, to be carried out almost without the aid of instruments. Until the time of Hipparchus himself, who was responsible for improving if not actually inventing the sighting-instrument known as the dioptra, the only optical aids were of the most primitive kind, such as the gnomon or vertical rod and the *polos* or sun-dial.

Yet the chief importance of fourth-century astronomy lies not in the advance in observational methods nor in the collection of empirical data, but in the example it provided of the successful application of mathematical methods to the study of complex natural phenomena. In part the impetus to apply such methods may be said to have come from philosophy. It was Plato who first insisted on treating astronomy as an exact science, doing so largely for reasons connected with his general epistemological theory and his conception of the relative worth of abstract thought and sensation, and it is hardly a mere coincidence that both Eudoxus and Heraclides were associated with Plato. It was, however, thanks to

97

the mathematical genius of Eudoxus that this new approach to astronomy became so influential. The geometry of the hippopede was sophisticated, and it seemed to show how extremely complicated phenomena might be reduced to simple regular movements. Thereafter astronomers differed in the models they proposed, but all were agreed in their methods and their aim: argument revolved round the advantages and disadvantages of different types of model in saving the phenomena, but that *some geometrical* model would provide the solution to the problem of celestial motion was assumed by all. Moreover, the work of the astronomers had a profound influence far beyond the field of astronomy itself, notably on cosmology. The success of the astronomers in making sense of the apparent irregularities in the movements of the heavenly bodies both stimulated and appeared to vindicate those who believed—as most ancient philosophers and scientists alike believed—that the world as a whole is the product of rational design.

8

Aristotle

FOR more than two thousand years, from the fourth century B.C. right down to the seventeenth century A.D., Aristotle exercised an unprecedented and unparalleled ascendancy over European science and cosmology. This very fact constitutes an obstacle to assessing his thought, which has often been misinterpreted through a failure to distinguish between Aristotle's own ideas and problems and those of his followers—between Aristotle himself and Aristotelianism. It is especially important, in his case, to consider his work first in relation to contemporary scientific problems and secondly in the light of what he himself had to say on the aims of the inquiry. His texts provide us with our most extensive evidence concerning the views of one ancient scientist on the value, purpose and methods of the inquiry into nature, and indeed he is just as important for his ideas on those topics as for his specific theories and discoveries.

Aristotle develops his doctrine of knowledge in the logical treatises known collectively as the *Organon* and in particular in the *Posterior Analytics*. There the regular Greek term for knowledge, *episteme*, is given a precise, technical meaning. *Episteme* is knowledge 'when we know the cause on which the fact depends as the cause of the fact, and that the fact could not be otherwise' (71b 10 *ff*). This knowledge is produced by demonstration, *apodeixis*, which is in itself a form of syllogism.[1] Being syllogistic, demonstration proceeds from premisses: the primary premisses from which

---

[1] A syllogism, in Aristotle, consists of two premisses and a conclusion and involves a total of three terms designating classes. The premisses and conclusion are linked either in the form of an inference (e.g. 'All broad-leaved trees are deciduous; all vines are broad-leaved trees; *therefore* all vines are deciduous') or—more

demonstration proceeds must themselves be indemonstrable, but known to be true, and he distinguishes between three types of such primary premisses, (*i*) axioms, (*ii*) definitions, and (*iii*) hypotheses. The axioms are the principles without which reasoning is not possible, such as the axiom that when equals are subtracted from equals, equals remain, and they are common to all the sciences. But definitions—that is, the assumptions of the meanings of terms—and hypotheses—the assumptions of the existence of certain things corresponding to those terms—are different for different sciences since they relate to the subject-matter of the particular science in question. In geometry, for example, both the meaning and the existence of points and lines are assumed, but everything else—such as the figures constructed from them—has to be proved to exist.

Knowledge in the strict sense is of things that cannot be otherwise than they are. It demonstrates connections that are necessary, eternal and 'universal' in a special sense that Aristotle explains. There is a 'universal' connection between subject and attribute when (*i*) the attribute is proved true of any instance of the subject, and (*ii*) the subject is the widest class of which it is proved true. The fact that the sum of the angles is equal to two right angles is a 'universal' attribute of triangle, but not of figure, nor of isosceles triangle. It is not a 'universal' attribute of figure, obviously, since it can only be proved true of some figures. But it is not a 'universal' attribute of isosceles triangle in the required sense either, for although it satisfies the first condition, it does not satisfy the second: while it can be proved true of all isosceles triangles, it is also true of triangles that are not isosceles.

The *Organon* is important for the fundamental contributions it makes towards the understanding of the structure of an axiomatic, deductive system. Aristotle

often—in the form of an implication: '*if* all broad-leaved trees are deciduous, and all vines are broad-leaved trees, *then* all vines are deciduous.'

carried the study of the conditions of proof further than any previous writer, and he undertook the first systematic analysis of deductive argument. Like Plato, he held that knowledge in the strict sense is irrefutable, and, as in Plato, we find mathematical examples used extensively to illustrate this conception. Thus in his discussion of demonstration in the first book of the *Posterior Analytics* almost all the examples are drawn—naturally enough—either from mathematics itself or from such mathematical sciences as optics, harmonics and astronomy. He has comparatively little to say, in the *Organon*, about induction, and in the only extended discussion of it (*Prior Analytics II*, Chapter 23) he suggests it may be reduced to a mode of deduction: he shows that when the induction is perfect or complete— that is when all the members of the class in question have been passed under review—the induction may be expressed in the form of a syllogism.

In the logical treatises Aristotle is chiefly concerned with deductive argument and with proof. But he also draws attention, as he does elsewhere in his work, to the distinction between the method to be used in demontration and the method of discovery or learning. In the former, he says, the starting-point is the universal and what is better known 'absolutely', while in the process of discovery or learning the starting-point is what is better known 'to us', that is, roughly speaking, the particulars or the immediate data of experience. Both these methods are relevant to the natural scientist, and indeed the latter is often just as important as the former. The main business of the physicist—as Aristotle's own practice bears out—is often not to present his arguments in the form of syllogisms which will make clear that the conclusions have been validly derived from the premisses, so much as to discover the causes themselves that form the middle terms of those syllogisms.

In practice Aristotle's method of procedure in the physical treatises is a complex one, and syllogistic reasoning plays a less prominent part than one might have expected from the importance attached to it in the *Organon*. His actual method varies from one branch of natural science to another, and from one problem to another, but some of the main recurrent features should be noted briefly. First, the subject-matter to be discussed must be defined. The way in which the problem is formulated is, indeed, often of crucial importance for Aristotle's theory. Thus, in the second book of *On Coming-to-be and Passing-away*, when he raises the question of the ultimate constituents of matter, he says that his inquiry is directed to finding the principles of perceptible body, and he goes on to suggest that these are the two pairs of contrary qualities, hot and cold, and dry and wet (see below pp. 107 *f*). He rejects out of hand the type of physical theory that the atomists and Plato had advocated, in which the differences between substances are ultimately reduced to quantitative, mathematical differentiae, and one of his main objections to such theories is that they mistake the nature of the problem, which is one of physics, not one of mathematics: the principles of perceptible body must themselves be perceptible contrary qualities.

In deciding what the problems are, he generally starts by considering the difficulties (*aporiai*) that are presented by what other theorists have suggested or by what is commonly accepted as true on a particular topic (the *endoxa*). This leads him to examine, sometimes at great length, the opinions that earlier thinkers have put forward. These surveys, which form such a remarkable feature of so many of Aristotle's works, are not undertaken for any purely historical motive, that is, to provide an accurate and exhaustive account of his predecessors' views: rather, his primary concern is always the solution of substantive problems, and earlier opinions are noted to help formulate the difficulties that have to be dealt with.

Once the difficulties have been stated and the common views outlined, Aristotle proceeds to his own solution. The nature of the argument varies according to the type of problem he is tackling, but broadly speaking his arguments fall into two main groups, which he himself distinguishes on many occasions, that is (*i*) dialectical, and (*ii*) empirical, or in Aristotle's usual terms, appeals to *logoi*, and appeals to *erga*. Under the first head we may include destructive arguments in which he refutes an opponent's view by posing a dilemma or by a *reductio ad absurdum*. Constructively, his analysis of many problems begins with a survey of all the theoretically possible alternatives, after which he proceeds to suggest the most satisfactory solution by a process of elimination. Often, when the difficulty under consideration itself takes the form of a dilemma, he suggests a way out by drawing a distinction. The careful definition of key terms and the analysis of their different senses form important parts of Aristotle's technique of argument in physics as in other inquiries. We find an example of this in his treatment of the problem of change. Here the dilemma had been posed: how can anything come to be, for it cannot come to be from what is not (for that is totally non-existent) or from what is (for then it exists already, and does not come to be). One argument that Aristotle uses rests on his distinction between potentiality and actuality. He suggests that a thing may come to be from what *is* in one sense, but in another sense *is not*, what it eventually becomes. The seed, for instance, *is* a tree in one sense—it is potentially a tree—although in another sense, of course, it *is not*—it is not actually one.

Apart from abstract, dialectical arguments, Aristotle often appeals to what he refers to as the 'facts' or the 'data' or the 'phenomena' (*erga, hyparchonta, phainomena*), though in interpreting these expressions we must be on our guard. An analysis of what he counts as evidence on such occasions shows that he includes a good deal besides what we should call empirical factors. Thus the term *phainomena* covers not only what we should call the phenomena, but also the accepted views, or what is usually said or thought about a subject—the 'appearances' in the sense of what appears to be the case. On the other hand these terms can and do refer to empirical evidence, including data obtained for the first time by Aristotle himself in his own extensive researches in, for example, the biological sciences.

Aristotle's general views on the importance and value of the study of nature emerge from his discussion of the good life in the tenth book of the *Nicomachean Ethics*. There he points out that the highest faculty that man possesses is reason, *nous*, and so the supreme activity of which he is capable is 'contemplation', *theoria*, which may be taken to include not only 'first philosophy' (that is, metaphysics) and mathematics, but also 'second philosophy' or *physike*: this last is defined as the study of natural objects that have a capacity for change or movement in themselves, and so it comprises not only the sciences we should call physics, chemistry and mechanics, but also the various branches of biology.

But in addition to texts in the *Ethics* and elsewhere in which he praises the life of reason in general terms, an extended passage in the first book of the work *On the Parts of Animals* is especially valuable for the light it throws on both the aims and the methods of one of the most important sub-divisions of the study of nature, namely that of zoology. In Chapter 5 he explains both how and why animals should be studied. He is on the defensive, anticipating and countering the criticisms of his contemporaries: the Platonists, especially, were no doubt profoundly shocked by his views concerning the value of observation and the importance he attached to studying particulars belonging to the world of becoming. Moreover Aristotle insists that observation of the external parts of animals is not enough, but must be supplemented by the use of dissection. He acknowledges (645 a 28 *ff*) that 'it is not possible to look at the constituent parts of human beings, such as blood, flesh, bones, blood-vessels and the like, without considerable distaste', but he maintains (644 b 29 *ff*) that 'anyone who is willing to take sufficient trouble' can learn a great deal about each of the different kinds of animals and plants.

The method that Aristotle advocates includes not merely observation, but also deliberate research, but he does not recommend 'pure' research in the sense of research carried out simply for its own sake. The aim of the study is to reveal the causes of things:

> For even in those kinds [of animals] that are not attractive to the senses, yet to the intellect the craftsmanship of nature provides extraordinary pleasures for those who can recognise the causes in things and who are naturally inclined to philosophy (645 a 7 *ff*).

And again:

> We should approach the investigation of every kind of animal without being ashamed, since in each one of them there is something natural and something beautiful. The absence of chance and the serving of ends are found in the works of nature especially. And the end, for the sake of which a thing has been constructed or has come to be, belongs to what is beautiful (645 a 21 *ff*).

The aim, and justification, of natural science is to reveal the causes that are responsible for the phenomena, and to understand Aristotle's view of 'physics' it is essential to be clear on his conception of 'cause'. He believes that four factors must be considered in giving an account of any object or event, whether natural or artificial. To give an account of a table, for instance, we must describe (*i*) its matter—for the table is made out of something, usually wood; (*ii*) its form—for the table is not just any lump of wood, but wood with a certain shape; (*iii*) its moving cause—for the table was made by someone, the carpenter; and (*iv*) its final cause—for when the carpenter made the table he made it for a purpose, to provide a flat raised surface which can be used to write upon or eat at. A similar analysis applies to natural objects too. Take the reproduction of a species of animals, say man. The matter, Aristotle believes, is supplied by the female parent, the mother. The form is the specific character of man and what marks him out from other animals: in Aristotle's usual definition, man is a rational two-legged animal. The moving cause is provided by the male parent, the father, and the final cause is the end towards which the process is directed, the perfect, fully grown man into whom the child will grow.

It is easy to see that Aristotle's conception of a 'cause' is a good deal wider than our own. Of the four factors he describes as causes in the two spheres of nature and art, only the moving cause and sometimes the final cause are causes in a sense we should recognise, and Aristotle's 'physics' correspondingly comprehends much more than the discussion of mechanical causation. But while he applies the same analysis of causation to both natural objects and artefacts, he recognises certain differences in the way in which it applies in these two fields. The most important difference concerns the final cause, for whereas in artificial production this is supplied by the conscious deliberation of the craftsman or artist, Aristotle denies that any *conscious* purpose is at work in nature. Nature does not deliberate. But that is not to say there are no 'ends' to natural processes. There are, but they are immanent in the objects themselves, in the living and growing animal or plant. Thus the child naturally grows into the mature man. The seed of the tree naturally grows into the mature specimen. It is true that something may intervene to prevent this, and Aristotle allows for this when he says that natural processes take place 'always *or for the most part'*. But if nature does not fulfil its ends always, as an absolute rule, it does so, he believes, in the vast majority of cases. Natural species reproduce according to kind; the young of each species naturally develop into mature specimens; even the inanimate elements, such as earth or fire, behave regularly in that they always fall or rise, when nothing impedes their movement. Nature as a whole is not random or haphazard, but exhibits order and regularity, and this order and regularity are the chief reason for Aristotle speaking of the 'ends' towards which natural processes are directed.

Aristotle's teleology has, then, certain distinctive characteristics. First, he makes it plain that he postulates no divine mind controlling natural changes from the outside. Secondly, he acknowledges that there are exceptions to the general rule that nature achieves its ends. Thirdly, his study of the ends of natural processes is in addition to, not to the exclusion of, his study of other types of causes, material, formal and moving. He investigates not only 'that for the sake of which' a natural process takes place, but also how it takes place, including what we should call mechanical causation. Fourthly, his interest in final causes is a particularly prominent feature of his biology, and here the study of ends is often a study of function: in this context his formal and final causes correspond, in many cases, to the structure and function of the part or organ.

No more than the briefest discussion of Aristotle's main physical and biological theories is possible here. At first sight his theory of the ultimate constituents of matter appears disappointingly retrograde. After the quantitative, mathematical theories of the atomists and Plato, Aristotle reverted to a qualitative doctrine. All other substances are thought of as compounds of the four simple bodies, earth, water, air and fire, and each of these in turn is treated as a combination of two of the four primary opposites: earth is cold and dry, water cold and wet, air hot and wet, and fire hot and dry, although it should be explained that the Greek terms *hygron* and *xeron* are wider than 'wet' and 'dry' in English, for *hygron* refers to both liquids and gases, and *xeron* especially, but not exclusively, to solids.

This theory owes a good deal to earlier doctrines, but what led Aristotle to propose it? The problem, as he states it in *On Coming-to-be and Passing-away* (329 b 7 *ff*), is to find 'the principles of perceptible body, that is of tangible body'. Every instance of a tangible quality can be represented as a position on a range or scale, and so the quality itself can be analysed in terms of the two opposite extremes, for example hard and soft, rough and smooth, coarse and fine. But some of these opposites can be derived from, or reduced to, others, for example the hard and the soft can be treated as modifications of the dry and the wet respectively. The minimum number of pairs from which all the tangible qualities can be derived is two, that is hot-cold and dry-wet, and the four possible combinations of these two pairs give the four simple bodies, earth, water, air and fire.

Aristotle conceives the problem as being to account for the sensible qualities of physical objects: his very way of stating the question commits him to a qualitative theory. To suggest that these qualities are in turn derived from more fundamental quantitative differentiae would, in his view, be to give the wrong sort of answer to the problem, indeed to mistake the nature of the problem itself. Moreover, granted that atomism was to prove more fruitful than any qualitative theory of matter, in the short term the doctrine that Aristotle proposed may well have seemed more promising. Certainly the accounts he gave of the ultimate constituents of matter and of the changes affecting the simple bodies stayed closer to what could actually be observed. It is obviously true that any physical object may be said to be either hot or cold and either dry or wet, whereas to associate the physical properties of substances with geometrical shapes must have appeared much more arbitrary. Again Aristotle could and did offer plausible-seeming interpretations of the changes affecting earth, water, air and fire. Take what happens when water evaporates or is boiled, becoming, on the Greek view, 'air'. Aristotle explained this as a change from the cold and the wet to the hot and the wet, that is an exchange of hot for cold. Conversely the opposite change, when 'air' condenses again to become water, was explained as a substitution of cold for hot, and again this type of theory provided a more direct explanation of the phenomena than any mathematical account of these changes.

It is hardly surprising that Aristotle's theory was more successful than its main rivals in antiquity. In particular it had the edge over any version of atomism as a working hypothesis in the investigation of the constitution of natural substances. Aristotle appears to have begun such an inquiry himself. In the fourth book of the *Meteorology* he discusses the physical properties of a variety of natural substances at some length. He considers, for example, which substances are combustible, which incombustible, which can be melted, which solidified, which are soluble in water or other liquids, and so on. He classifies natural substances broadly according to which of the simple bodies predominates in them, suggesting, for instance, that those substances that solidify in cold, but are dissoluble in fire, consist mainly of water, while those that are solidified by fire are composed predominantly of earth. The actual conclusions drawn are, as these examples show, naïve, and he attempts no precise estimate of the proportions of the simple bodies in the various compounds he considers. Even so the fourth book of the *Meteorology* is the first important attempt, in antiquity, to begin the incredibly complex task of collecting and collating information about the physical properties of natural substances and their reactions to certain simple tests.

According to Aristotle the matter of which everything on earth is composed is earth, water, air and fire, but he maintains that the heavenly bodies consist of a quite different substance, a fifth element, *aither*. This doctrine has probably attracted more scorn and derision than any other would-be scientific theory from antiquity, so it is particularly important to understand Aristotle's motives for suggesting it. The problem, as he saw it, was to account for the eternal, unvarying, circular movements of the heavenly bodies. That their movements are unvarying he believed to have been established by observation. He claims some acquaintance with the work of Egyptian and Babylonian, as well as Greek, astronomers, and he remarks that no variation in the outermost heaven or any of its parts has ever been recorded. The changes in the positions of the fixed stars are obviously quite regular, and the apparently irregular paths of the planets were—as we have seen in Chapter 7—assumed to be reducible to combinations of regular, circular movements.

But how could this supposed fact of the eternal, unvarying movements of the heavenly bodies be explained? The natural movement of the four terrestrial elements is either upwards or downwards, from or to the centre of the earth: fire and air naturally rise, water and earth naturally fall, when nothing impedes their movement. They can, of course, be moved in other directions as well, as when a heavy object such as a stone is hurled into the air. But such a movement is not natural, but enforced: it requires a propelling agent, unlike the natural movement of a flame rising or of a heavy object falling. But the circular motion of the heavenly bodies, being eternal, cannot be enforced. It must, then, be natural. Yet an object that moves naturally in a circle cannot, he argued, be one of the terrestrial elements or a compound of them. Their natural movements are upwards or downwards, and if they are moved in a circle, as when a stone is whirled round in a sling, this motion is, in part at least, enforced. And so there must be something, a fifth element, which moves naturally and continuously in a circle.

But while this is the main theoretical argument which led him to the conclusion that there must be a fifth element, other factors, some of them empirical ones, also influenced his doctrine. He had some appreciation of the extreme distance of the heavenly bodies from the earth and of the vast expanse of the heavens compared with the volume of the earth and its surrounding atmosphere. One of the arguments used to support his doctrine of the fifth element is that if either air or fire, for example, had been the constituent of the vast space between the earth and the outermost stars, the earth itself would long ago have been destroyed. The four terrestrial elements are each either hot or cold, and either dry or wet, and for these to continue to exist there must be an approximate balance between them. But the immense space of the heavenly region must, then, be filled by some other element which is not characterised by these opposite qualities, for otherwise the sublunary elements would be destroyed.

Aristotle himself points out that his doctrine of the fifth element fits in with traditional Greek religious beliefs in the divinity of the heavenly region. But apart from religious considerations, the doctrine was an attempt to deal with a serious physical problem, the continuous circular movement of the heavenly bodies, and the abstract arguments and empirical evidence brought in its support are by no means negligible. It left, however, several difficulties unresolved.

First, he offered no account of the join between the celestial and the terrestrial region. At or just below the sphere on which the moon is located the sublunary elements give way to the fifth element, *aither*. This has a quite different natural motion, and is neither hot nor cold, neither wet nor dry, and yet it must somehow transmit movement to the sublunary elements and indeed do so without itself in any way being affected by them. Secondly, how can the heavenly bodies emit light, and in the case of the sun heat, when—if they consist of *aither*—they cannot themselves be hot? Here Aristotle tentatively suggested that light and heat are produced by the friction of their movements, even though they themselves do not become hot. A third unsatisfactory feature of his theory concerns the position of the sun. We have seen (pp. 92 f) that in converting Eudoxus' model of concentric spheres from a purely mathematical into a mechanical system, he postulated certain reacting spheres the purpose of which was to counteract the movements of the next highest heavenly body. But Aristotle recognised that it is not the lowest heavenly body, the moon, but the second lowest, the sun, that is responsible for change in the sublunary region, particularly the variation in temperature associated with the cycle of the seasons. Yet how, we may ask, can this be the case, when the sun's movements are cancelled out by the reacting spheres between it and the moon?

There are, then, serious problems which Aristotle's doctrine of the relationship between the heavenly bodies and the sublunary region leaves unresolved. But although he was aware of some at least of these difficulties, they did not lead him to modify any of the essential features of his theory. The sun obviously is responsible for seasonal changes on earth, and the heavenly bodies as a whole evidently do emit light. Although they consist of *aither*, then, the higher heavenly bodies do, somehow or other, affect the sublunary region. But the doctrine of *aither* itself could not be abandoned without leaving a far greater difficulty unresolved, namely how to account for their continuous circular movement.

Further difficulties relate to Aristotle's doctrine of terrestrial motion, another highly influential, and much criticised, area of his physics. Once again we must judge his theories against the background of earlier speculation. It is hardly an exaggeration to say that before Aristotle there is nothing that can be called dynamics at all in Greek science. The Presocratics refer, in various contexts, to the principle that like things attract one another, but that generalisation covers a wide range of phenomena. The action of gravitational forces may be included under that heading, as when a heavy object such as a clod of earth 'seeks its like' in falling downwards, but so too may the behaviour of gregarious animals, which was an example actually used by Democritus to illustrate how like is drawn to and known by like (fragment 164). Aristotle's texts provide, then, the first general statements concerning the relations between the various factors governing the speed of a moving body. Yet Aristotle himself undertook no systematic account of the problems of dynamics, and the statements in question are made in various contexts in the physical treatises in the course of his discussion of such questions as the existence of a void or that of a body of infinite weight.

Take first some of his statements concerning natural motion, that is the motion of freely falling or freely rising bodies. At 273 b 30 *ff* in *On the Heavens*, for instance, he implies that the speed is directly proportional to the weight of the body:

> If a certain weight move a certain distance in a certain time, a greater weight will move the same distance in a shorter time, and the proportion which the weights bear to one another, the times too will bear to one another.[1]

Elsewhere we find him suggesting that the speed will be inversely proportional to the 'density' of the medium through which the movement takes place. In the *Physics* (215 b 4 *ff*) he considers movement through air and through water and says:

> by so much as air is thinner and more incorporeal than water, the object will move through the one faster than through the other.[2]

Then in enforced motion he implies in *Physics* VII, Chapter 5, for instance, that the speed is directly proportional to the force applied and inversely proportional to the weight of the body moved. Yet he recognised that in certain cases this rule does not apply. If a force A moves an object B a distance C in a time D, it does not necessarily follow that half the force A/2 will move the same object B half the distance C/2 in the same time D, since it may be the case that half the force is insufficient to move the object at all: otherwise, as he observes (250 a 17 *ff*)

> one man might move a ship, since both the motive power of the ship-haulers, and the distance that they all cause the

[1] From the Loeb translation by W. K. C. Guthrie (Cambridge, Mass., Harvard University Press; London, Heinemann, 1939).

[2] Based on the Oxford translation, *The Works of Aristotle translated into English*, edited by W. D. Ross (Oxford, Clarendon Press), *Physics*, R. P. Hardie and R. K. Gaye (Vol. II, 1930).

ship to traverse, are divisible into as many parts as there are men.

The general rules that his statements suggest are wide of the mark. Yet they are less at variance with observed phenomena than might at first be supposed, in view of the differences between Aristotelian and Newtonian dynamics. He has often been condemned for assuming that the speed of a freely falling body varies directly with its weight. But the fact is that in air heavier bodies do fall more rapidly than lighter ones of the same shape and size, although this is not true in a vacuum. He was correct in assuming that there is some relationship between weight and speed in motion that takes place through a medium, although the relationship is not a simple one of direct proportion. Similarly it is obviously true that motion through a dense medium is generally slower than through a rare one, but again he oversimplified the relationship in treating it as one of a direct proportion.

The main shortcoming of Aristotle's dynamics is not so much a failure to pay attention to the data of experience, as a failure to carry abstraction far enough. He saw that certain factors, such as shape, have to be discounted in trying to establish the laws governing the speed of moving objects. Yet while we should include among the factors that have to be so discounted the effects of the resistance of the medium, Aristotle assumed that motion must take place through a medium. Indeed because he believed that speed is inversely proportional to the density of the medium, he denied that motion through a void is possible, since the speed would have to be infinitely great—he concluded, accordingly, that a void cannot ever exist in reality. But in thus assuming that motion necessarily takes place through a medium, he may be said to have stayed too close, rather than not close enough, to the data of experience. The paradigms of motion in his dynamics are such obvious—yet as we now know extraordinarily complex—cases as a ship being hauled through the water: the ship can be hauled more easily when it is unladen, its speed increases when the number of men hauling it is increased and so on. But the paradigm of Newtonian dynamics is one that we never observe except in artificial conditions, namely frictionless movement through a void.

It is, however, true both that Aristotle failed to carry out certain simple tests that would have indicated the inaccuracy of some of his propositions, and that there are certain logical inconsistencies between the general rules that his statements imply that might have suggested to him that he had oversimplified certain problems. Later theorists were indeed to criticise Aristotle's doctrine both on abstract and on empirical grounds: in the sixth century A.D. Philoponus, for example, provided experimental evidence to refute the doctrine that the speed of a falling body is directly proportional to its weight. But while there are certainly grave inadequacies in Aristotle's dynamics, it may be repeated that he was the first in the field, and if there is a lesson to be learned from this part of his physics for his method and approach to scientific problems as a whole, it is not that he blandly ignored the observed facts in constructing theories on *a priori* principles, but rather that his theories are hasty generalisations based on rather superficial observations.

The branch of natural science that received most attention from Aristotle is biology—the biological treatises make up more than a fifth of his total extant work—and the reason for this is clear. Living creatures and their parts provided far more evidence of the roles of form and the final cause than inanimate objects did. As we have seen (pp. 104 f), he felt the need to justify the study of animals and was conscious of being a pioneer in this field. Against the Platonists and all who disparaged the use of observation, he insisted on the value and importance of detailed research in biology. ' Anyone who is willing to take sufficient trouble' can learn a great deal concerning each one of the different kinds of animals and plants, and 'to the intellect the craftsmanship of nature provides extraordinary pleasures for those who can recognise the causes in things'.

The range of his zoological researches is remarkable. Well over 500 different species of animals, including about 120 kinds of fish and sixty kinds of insects, are referred to in the biological works. His data were collected from a wide variety of sources: he relied a good deal on fishermen, hunters, horse-trainers, bee-keepers and the like, but also undertook his own personal researches. In some cases we can infer with some probability when and where his work was carried out. The biological treatises contain some particularly detailed accounts of the marine animals in the lagoon of Pyrrha on Lesbos and we know that Aristotle spent a couple of years (344–342) on that island. While he was certainly not the first biologist to use the method of dissection, he was the first to do so extensively. It is impossible to give a precise estimate of the number of species he dissected, and they evidently did not include man: in the *Inquiry Concerning Animals* (494 b 22 *ff*) he remarks that 'the inner parts of man are for the most part unknown, and so we must refer to the parts of other animals which those of man resemble, and examine them'. Nevertheless our texts contain many detailed reports that give information that could only have been obtained from dissections, and numerous passages refer directly to the method. At 496 a 9 *ff* in the *Inquiry Concerning Animals*, for instance, he says that

in all animals alike . . . the apex of the heart points forwards, although this may very likely escape notice because of a change of position while they are being dissected.

Again in his account of the male generative organs of viviparous land-animals in general he remarks that the membrane now known as the *tunica vaginalis* must be cut to reveal the relation between the ducts it encloses:

The ducts that bend back again and those that lie alongside the testicle are enclosed in one and the same membrane, so that they appear to be one duct, unless the membrane is cut open (*Inquiry Concerning Animals*, 510 a 21 *ff*).

To be sure, the biological treatises contain, as Aristotle's critics have been quick to point out, many mistakes, some of them simple ones—as when he gives the number of teeth in women or the number of ribs in man incorrectly—others more serious, such as his belief that the brain is bloodless, and the influential doctrine associated with this, that the heart is the seat of sensation. Yet his attitude towards the evidence he collected from his informants is, in general, a cautious and critical one. He often speaks of the need to verify the data—particularly concerning rare animals or exceptional phenomena—and, when the available evidence appears to him to be inadequate, he draws attention to this fact. Two passages will illustrate this. In *On the Generation of Animals* (741 a 32 *ff*) he considers the possibility of some species of animals reproducing parthenogenetically:

> If there is any kind of animal which is female and has no separate male, it is possible that this generates offspring from itself. Up till now, at least, this has not been reliably observed, but some cases in the class of fishes make us hesitate. Thus no male of the fish called *erythrinos* has ever been seen, but females have, including females full of roe. But of this we have as yet no reliable proof.

Again, after a highly sententious discussion of the problems connected with the generation of bees in book III, Chapter 10 of the same work, he concludes by acknowledging the inadequacy of the data available:

> This, then, seems to be what happens with regard to the generation of bees, judging from theory and from what are thought to be the facts about them. However, the facts have not been sufficiently ascertained. And if they ever are ascertained, then we must trust the evidence of the senses

117

rather than theories, and theories as well, so long as their results agree with what is observed (760 b 27 *ff*).

Some of the discoveries that Aristotle made or recorded are justly famous. One of the most remarkable is his account of a species of dog-fish, the so-called 'smooth shark' (mentioned above, on p. 18, in connection with Anaximander). The species in question, *Mustelus laevis*, is externally viviparous, as indeed are several of the cartilaginous fishes, but it is exceptional in that the embryo is attached by a navel-string to a placenta-like structure in the womb of the female parent. Aristotle's description in the *Inquiry Concerning Animals* (VI, Chapter 10, 565 b 1 *ff*) is clear and precise, yet it was generally disbelieved until Johannes Müller published the results of his investigations of this and related species in 1842—investigations which largely vindicated the accuracy of Aristotle's account. And it is not only for the discovery of such exceptional phenomena as this that Aristotle has won praise from naturalists, but also for his meticulous descriptions of the external and internal parts of such familiar species as the crawfish.

Aristotle demonstrates his skill as an observer in passage after passage in the biological treatises. But his chief motive for studying animals was, as we have seen, not description, but explanation, to establish the causes at work and especially the formal and final causes. The work *On the Parts of Animals* deals mainly with the causes of the various parts of the body, and in *On the Generation of Animals, On the Motion of Animals, On the Progression of Animals* and the short treatises known as the *Parva Naturalia* he also tackles a wide range of physiological problems, including nutrition and growth, respiration, locomotion and, especially, reproduction. Not surprisingly, the positive conclusions that he reached on these obscure questions are usually wide of the mark. Nevertheless his discussions have at least two considerable merits, first, in the clarity with which the problems themselves are formulated, and secondly, in the ingenuity and acuteness with which he develops and analyses the arguments on either side.

One example which will illustrate this is his discussion of one of the fundamental problems of reproduction, the question of whether the seed is drawn from the whole of the parent's body or not. The view that it is, the 'pangenesis' theory, had been advocated by the atomists and by some of the medical writers (see p. 63), but is severely criticised by Aristotle. In *On the Generation of Animals* (I, Chapters 17 and 18) he sets out the problem and cites the main evidence and arguments that had been used to support pangenesis. So far as the evidence goes, he questions or flatly denies its validity. Thus it had commonly been supposed that not only congenital but also acquired characteristics—Aristotle's terms are *symphytos* and *epiktetos*—are inherited, and that mutilated parents, for example, have mutilated offspring. But to this Aristotle replied by simply denying that this is always the case.

Against pangenesis he brings some ingenious and telling counter-arguments. One aims to show that the theory is incoherent by posing a dilemma. The seed must be drawn either (*i*) from all the uniform parts—Aristotle means flesh, bone, sinew and so on—or (*ii*) from all the non-uniform parts—by which he means the hand, the face and so on, or (*iii*) from both. But against (*i*) he objects that the resemblances that children show to their parents lie rather in such features as their faces and hands than in their flesh and bones as such. And against (*ii*) he points out that the non-uniform parts are actually composed of the uniform ones. A hand is made up of flesh, bone, blood, nail and so on. Against (*iii*) he uses the same consideration. Resemblances in the non-uniform parts must be due either to the material—but *that* is simply the uniform parts—or to the way in which the material is arranged. But if to the latter, nothing can be said to be 'drawn' from the *arrangement* to the seed, for the arrangement is not itself a material factor. In either case the seed cannot be drawn from such parts as the hands or face, but only from what those parts are made of. But then the theory loses its point, which was that all the individual parts of the body, and not merely all the constituent substances, supply material to the seed.

He rejects the pangenesis doctrine, then, and in the main he was right to do so, even though his own positive theory concerning what each parent contributes to the offspring was in certain respects very mistaken. He believed, for example, that the semen of the male contributes no material to the embryo, but merely supplies the form and the efficient cause of generation.

A second more fundamental controversy in biology concerned the role of the final cause itself. Whereas both Plato and Aristotle insisted on the element of rational design throughout nature and in living creatures in particular, other theorists, and especially Empedocles and the atomists, had generally adopted a mechanistic, non-teleological stand-point in their accounts of natural causation. The evidence about Empedocles is particularly interesting, though extremely obscure. In *On the Parts of Animals* (640 a 19 *ff*) Aristotle ascribes to him the view that

> animals have many characteristics that are the result of incidental occurrences in their formation,—for instance the backbone is as it is [divided into vertebrae] because the foetus becomes contorted and so the backbone is broken.

Again Simplicius, dealing, in his *Commentary on Aristotle's Physics* (371, 33 *ff*), with the famous fragment in which Empedocles spoke of the birth of 'man-headed oxen' (fragment 61), reports that he held that

> during the rule of Love there came to be by chance first of all the parts of animals, such as heads and hands and feet, and then these 'man-headed oxen' came together, and 'conversely there sprang up' ox-headed men . . . . And as many of these as were fitted together to one another so as to ensure their preservation, became animals and sur-

vived. . . . For all that did not come together according to
the proper formula (*logos*) perished.

Although there are obvious superficial similarities
between these notions and the doctrine of the evolution
of species, it must be remembered first, that Empedocles
was not attempting a systematic account of the origin of
natural species at all, and secondly, that his ideas were
developed in the context of a highly fanciful cosmological doctrine of the cycle during which the two cosmic
forces of Love and Strife come to rule in turn.

In arguing against those who tended to deny design
in living creatures, Aristotle no doubt believed that the
evidence pointed overwhelmingly in his favour. He
knew, of course, that abnormalities and monstrous births
do occur, but the important point, in his view, was that
these were the exceptions to a rule that held good in the
vast majority of instances. One of the chief considerations that he brings against Empedocles and the atomists
is simply that natural species reproduce according to
kind. In *On the Parts of Animals* (640 a 22 *ff*) he says
that Empedocles ignored the fact that the seed which
produces any animal must have the appropriate specific
character of that animal. It is a man that begets a man,
an ox that begets an ox: the idea that natural species
themselves were the result of chance mutations would
have seemed to him not only to have no direct evidence
in its support, but also to fly in the face of the existing
evidence of the normal circumstances of the reproduction of natural species.

The notions of form and final cause permeate the
whole of Aristotle's philosophy. They are fundamental
not only to his natural science, but also to his cosmology:
the primary cause on which the universe depends and
from which all movement is ultimately derived is an
Unmoved Mover which is said to bring about movement
as final cause, as the good that is the object of desire and
love. Form and finality are equally prominent in his
ethics and politics too, for his ideas of the good life, and

of the good state, are based on his conception of man's proper ends or function. Man has a unique place on the scale of being: he shares with the gods the possession of reason, but with the other animals the possession of his other vital faculties, such as sensation, nutrition and reproduction. At the same time it is a *single* scale of being that comprehends gods, men, animals, plants and inanimate objects. Different natural objects have different forms and ends, but every kind of natural object from the divine heavenly bodies down to the humblest pebble seeks and aspires to the form and end appropriate to it.

The very comprehensiveness of Aristotle's philosophy was a major factor that contributed to its enormous influence in antiquity: the doctrine of causes laid down both what types of question to ask and the terms in which to answer them. Yet his immediate followers were far from being merely slavish imitators of his ideas. The next two heads of his school, the Lyceum,[1] were Theophrastus of Eresus (in Lesbos) and Strato of Lampsacus, both of whom were original thinkers of considerable calibre, and both of whom criticised and rejected parts of Aristotle's teaching. Thus Theophrastus criticised the doctrine of the final cause in his *Metaphysics*, and in his treatise *On Fire* he raised doubts about whether fire is on a par with the other three simple bodies.

The history of the detailed scientific theories of Theophrastus, Strato and the other members of the Lyceum falls outside the scope of this study, but from one point of view the work of the school as a whole is relevant to the assessment of Aristotle himself. The idea of collaboration in research owes something to such earlier groups as the Pythagorean communities, the medical schools

[1] The Lyceum was a grove just outside Athens where other teachers besides Aristotle gave lectures, but the name came to be attached to Aristotle's school in particular. Although he began teaching there shortly after his return to Athens in 335, it was probably not until after his death in 322 that the school acquired its own property in the Lyceum and had, like Plato's Academy, the legal status of a *thiasos* or religious association.

and Plato's Academy. But the researches undertaken by Aristotle and his colleagues and pupils far surpassed anything that had been contemplated, let alone achieved, before. First, there was a series of histories of different branches of speculative thought: these may be seen as a natural development from the outline accounts of the views of his predecessors that Aristotle himself gave in such treatises as the first book of the *Metaphysics*. Thus Theophrastus undertook the histories of the main physical doctrines and of the theories of sense-perception in earlier thought, Meno[1] the history of medicine, and Eudemus the histories of geometry and astronomy.

Secondly, there was research in the social sciences. Here the most notable work was the series of 158 constitutional studies of which the *Constitution of Athens* is the sole surviving example: while Aristotle planned the series as a whole, he probably composed no more than a small proportion of these studies himself.

Thirdly, there was work in the natural sciences. Aristotle's own zoological treatises represent to some extent, indeed in the case of the *Inquiry Concerning Animals* to a large extent, the results of a joint effort in research. These zoological studies were complemented by the equally full botanical treatises of Theophrastus, the *Causes of Plants* and the *Inquiry Concerning Plants*. Then the study of the constitution of natural substances in the fourth book of the *Meteorology* was followed by a detailed investigation of minerals, again by Theophrastus, the treatise *On Stones*. Finally in dynamics Aristotle's own unsystematic discussion of movement and weight in the *Physics* and *On the Heavens* was followed by the work of Strato, who, according to Simplicius, undertook certain investigations connected particularly with the phenomena of acceleration.

The scale on which research was carried out in the Lyceum was unprecedented. Nor was this merely fortuitous, but the result of applying Aristotle's own

[1] Aristotle's pupil is not, of course, to be confused with the character in Plato's dialogue of the same name.

methodological principles, and in particular his insistence on reviewing the 'data' and the 'common opinions' both in order to discover the problems and as the first step towards resolving them. Much of Aristotle's work in natural science is coloured by fundamental assumptions that he shared with his master, Plato. Both philosophers believed that the world is the product of rational design. Both held that what the philosopher investigates is the form and the universal, not the particular and the accidental. Both considered that it was only certain and irrefutable knowledge that could be termed knowledge in the strictest sense. But despite these important similarities between Plato and Plato's most brilliant pupil, in other respects they disagreed profoundly on points that are reflected in their attitude towards natural science. Where Plato spoke of the Forms as existing independently of the particulars, Aristotle denied this, maintaining that while form and matter are distinguishable in thought, they are not distinguishable in fact in the objects in the world around us. The form of a table, for instance, does not have a separate existence: it does not exist, that is, except in conjunction with matter of a certain sort. Again where Plato, in insisting on the role of reason, had depreciated that of sensation, Aristotle reinstated observation. Both men made important contributions to what we may call the philosophy of science, but the nature of their contributions was very different. Whereas Plato was chiefly responsible for the idea of applying mathematics to the understanding of phenomena, one of Aristotle's fundamental and lasting contributions was that he both advocated in theory, and indeed demonstrated in practice, the value of undertaking detailed empirical investigations.

9

Conclusion

THE foregoing chapters have outlined the main ideas and problems of Greek science from its beginnings to Aristotle. It is now time to review the data we have collected and to raise, first, certain general questions concerning the social, economic and ideological setting of early Greek science, before attempting to sum up its character and assess its limitations and achievements. How far can we determine the motives that prompted men to engage in scientific research in antiquity? What were the economic factors at work—that is, how were scientists financed or how did they earn a living? These are difficult questions and the evidence hardly allows us to give definite answers to them. Nevertheless certain fundamental points are clear.

First it is obvious that the men whose ideas we have been considering do not form a clearly defined group of 'scientists'. As I noted at the outset, there is no single term in Greek that is exactly equivalent to our 'science'. Even when we confine our attention to the period from Thales to Aristotle, the theories that fall under the heading of 'Greek science' are enormously diverse. The people who produced them were known to their contemporaries not as 'scientists' but as philosophers or *physikoi* or mathematicians or doctors or sophists. And even within those general groups there are further important differences between the attitudes of different individuals towards the inquiry they were engaged on.

Our questions concerning the motives that underlie scientific investigations and the economics of science are, then, much more complex than would have been the case had science as such been a recognised profession or career. Take first the economics of science. There are three possible answers to the question of the sources of

125

liveiihood of the men who conducted scientific research, these three being complementary to one another, not mutually exclusive, (*i*) independent means, (*ii*) practising a paid 'profession' such as medicine or teaching, and (*iii*) patronage.

Of these it would appear that the most important is the first, independent means. It is extremely difficult, to be sure, to estimate the wealth of the men we are interested in. Very few ancient writers speak about their financial circumstances, so for the most part we are reduced to basing our conjectures on the information to be gleaned from such secondary sources as Diogenes Laertius' *Lives of the Philosophers*. Sometimes, for example, the family of an important thinker is recorded. Some were the sons of artisans or craftsmen: we are told that Pythagoras was the son of a gem-engraver and Theophrastus the son of a fuller. Yet among the philosophers such men were probably the exceptions. In a large number of instances Diogenes speaks of the wealth of the early Greek philosophers. He does so concerning Heraclitus, Parmenides, Anaxagoras and Empedocles, for example. Admittedly this evidence must be treated with caution. Yet the idea that leisure and material prosperity are the preconditions of an interest in philosophy or natural science was already frequently expressed in the fourth century by such writers as Isocrates and Aristotle, and what Diogenes records is at least consistent with what we know of the general economic conditions of the Greek world in the fifth and fourth centuries B.C. Quite a number of the citizens of the larger city-states were evidently sufficiently free from immediate financial worries to be able to spend a good deal of time and energy on a wide range of relatively unproductive, or even counter-productive, activities, such as political intrigues and law-suits, as well as many other cultural pursuits besides the inquiry concerning nature.

Science as such was no profession. But important contributions were made to the development of Greek scientific thought by men who earned money from the practice of what may loosely be described as professions, especially medicine and teaching.[1] The conditions under which medicine was practised in the fifth and fourth centuries have been reviewed in Chapter 5. The doctor had no formal professional qualifications to cite when he wished to support a claim to be able to heal the sick but, as I noted, several of the medical writers insist on the difference between the experienced doctor and the layman, and again between the doctor and the quack. Despite the insecurities of medical practice, medicine provided many doctors with a livelihood and some with much more than a mere livelihood. The importance of a flourishing medical 'profession' for Greek science is obvious. Throughout antiquity notable contributions were made not only to the biological sciences, but also to the discussion of more general problems in physics, cosmology and the methodology of science, by men who were first and foremost medical practitioners. This is true not only in our period, but also later: the great third-century B.C. Alexandrian anatomists and physiologists Herophilus and Erasistratus, and the greatest of all ancient biologists, Galen of Pergamum (second century A.D.), were all medical men.

Teaching was or became important in the development of Greek science for two reasons. First, it too, like medicine, could provide a livelihood, and secondly, such institutions as the Academy and the Lyceum provided the opportunity for collaboration in research. Evidence concerning the earnings of the sophists comes mostly from hostile sources such as Plato. In the *Hippias Major* (282de), for instance, Hippias is made to boast that he earned more than 150 minae[2] from a short visit to Sicily. But even when we allow for some exaggeration in such

[1] A third such 'profession' which became important after our period is that of the *architekton*, this being the term used not only of architects and town-planners, but also of engineers and the designers of siege-engines and of weapons of war in general.

[2] See above, p. 53, footnote 1 on the approximate value of a mina in the fifth century.

statements, we can be sure that several of the fifth-century sophists, such as Protagoras, Hippias and Gorgias, amassed considerable fortunes. Admittedly not many of the subjects they lectured on have any direct bearing on science, but exceptions must be made of astronomy and geometry, both of which were taught by Hippias. While none of the most famous fifth-century sophists could be described as an original scientist, some of them helped to raise the general level of education on certain scientific subjects.

The foundation of such schools as the Academy and the Lyceum in the fourth century marks an important development. The extent to which the members of such institutions devoted their attention to scientific issues, and the nature of the issues that interested them, varied a great deal. The Academy was stronger in mathematics and weaker in the biological sciences, while the reverse was true of the Lyceum under Aristotle and Theophrastus, and in other fourth-century schools, such as that of Isocrates, science played no part at all. Again the financial arrangements differed from one school to another. Usually the pupils paid fees for attending lectures, although the Academy may have been an exception, at least at first: alternatively, or in addition, they were asked to make contributions towards the general upkeep of the school. It is, however, clear that teaching could be an important primary or supplementary source of income not only for those who taught such subjects as rhetoric and politics, but also for those whose main interests were in astronomy or physics or biology.

The extent and importance of patronage in our period are harder to gauge. Many of those who engaged in scientific inquiry appear to have enjoyed private fortunes. If that was so, then it seems likely that the relationship between scientists and their rich and powerful acquaintances was often simply one of friendship, not one of patronage in the sense that the scientist received direct financial support. Thus Anaxagoras was the friend and teacher of Pericles, but it would be rash to assume any significant financial element in their relationship. Anaxagoras himself, we are told, was born into a wealthy family, and whether he received any money for such instruction as he gave Pericles is open to doubt. Plato at least contrasts the wise men of Anaxagoras' time and earlier who did not teach for money, with later sophists, such as Protagoras, who did (*Hippias Major* 281c *ff*). In many of the stories that are recorded in antiquity both in and after our period concerning the dealings between philosophers or scientists on the one hand, and tyrants or kings on the other, it is the tyrant or king who is represented as making the overtures, and while many such stories should be treated with scepticism as presenting an interpretation unduly flattering to the philosopher or scientist, that does not apply to them all. Thus Plato did not solicit the invitations he received first from Dion and then from the younger Dionysius to visit Syracuse. No doubt Aristotle, whom Philip had chosen as tutor to Alexander in 342, enjoyed Macedonian approval and support both then and later when he returned to Athens in 335. Yet the extent to which he or his school benefited from direct subventions from Alexander or his regent Antipater may have been quite small, and the stories we read in Pliny and elsewhere of Alexander ordering large numbers of people throughout his empire to assist Aristotle in his zoological studies are, no doubt, apocryphal. It was not until the foundation of the Museum at Alexandria (c. 280 B.C.) that the patronage of powerful rulers became an important factor in furthering the work of Greek scientists. Earlier institutions such as the medical schools, the Academy and the Lyceum were entirely self-supporting or very largely so. In any case the requirements of scientists in our period were extremely modest: indeed throughout antiquity astronomers, physicists and biologists alike used only the simplest instruments and the most rudimentary apparatus in their work.

Although many aspects of our problem remain obscure, one general conclusion which applies to Greek science at any period is quite clear: we must discount any idea that there were major financial incentives to engage in scientific investigations. Certainly the more successful teachers and doctors made a good deal of money in the fifth and fourth, as also in later, centuries. We should not underestimate the part played by economic considerations in attracting recruits to the 'arts' of sophist and doctor. But that does not explain why some men within those two categories engaged in scientific research. Financial incentives may help us to understand why some men became doctors, but not why some men became scientific doctors, for example anatomists, physiologists or embryologists.

But if we have to discount direct financial incentives, as indeed we must, how far can we go towards answering our question of the motives for which scientific investigation was undertaken? If we review what ancient writers themselves have to say on this subject, the dominant theme, found in many variations, is undoubtedly that the inquiry concerning nature is its own reward. The most common justification offered for both philosophy and natural science is that knowledge is valuable for its own sake. The myth of the unworldly philosopher is already applied to Thales in the story which we find in Plato (*Theaetetus* 174a) of his falling down a well while contemplating the heavens—although, as we have already noted (p. 14), other stories concerning Thales present a very different picture of him, as the astute man of business, in the account of his making a corner in olive presses in Aristotle (*Politics* 1259 a 6 *ff*), or as an engineer, in the account of his diverting the river Halys in Herodotus (I, 75). Again some of our sources attribute to Pythagoras the distinction between the three lives, those of contemplation, honour and wealth, and the view that the best of the three is the first, the life of contemplation. In these two instances we depend on secondary sources for our evidence, but original texts that speak of the joys and merits of the life of wisdom begin before the end of the fifth century. Empedocles

130

(fragment 132) describes as fortunate, *olbios*,[1] the man who 'has gained the riches of divine intelligence', and in a well-known fragment (910) of a lost tragedy Euripides too used the same word 'fortunate' to describe the man who engages in inquiry (*historia*) and who 'observes the ageless order of immortal nature'.

Then Plato and Aristotle not only lend their considerable authority to this ideal, but also give it a rational basis in their psychology. Both agree that philosophy, as the activity of the highest part of the soul, the reasoning part, is essential to true happiness. For Plato, as we saw, the study of the changing world of becoming is inferior to the study of the immutable Forms, but the former inquiry is far from valueless since it reveals the intelligent ordering of the universe. For Aristotle, too, the life of 'contemplation' (*theoria*) is the supreme life, and he explicitly vindicates the study of biology on the grounds that 'there is something beautiful' in every species of animal. He develops his view of the relative worth of utilitarian and non-utilitarian studies in the first two chapters of the *Metaphysics*. At 981 b 13 *ff* he echoes the common Greek admiration for those who had invented arts or crafts:

> At first he who invented any art whatever that went beyond the common perceptions of man was naturally admired by men, not only because there was something useful in the inventions, but because he was thought wise and superior to the rest.

But he then goes on:

> But as more arts were invented, and some were directed to the necessities of life, others to recreation, the inventors of the latter were naturally always regarded as wiser than the inventors of the former, because their branches of knowledge did not aim at utility.

[1] The word is derived from *olbos*, the primary meaning of which is wealth, material prosperity.

Aristotle does not so much ignore the possibility of putting theoretical ideas to practical use, as positively glory in the ideal of the pursuit of knowledge for its own sake, and he shows what seems to us an almost incredible complacency concerning the material conditions of life in his day:

> Hence when all such inventions [that is, those aiming at utility] were already established, the branches of knowledge which do not aim at giving pleasure or at the necessities of life were discovered, and first in the places where men first began to have leisure.[1]

Although he goes on, in that passage, to speak of Egyptian advances in mathematics, he has a similar point to make concerning the origins of philosophy in Greece (982 b 12 *ff*):

> For it is owing to their wonder that men both now begin and at first began to philosophize . . . therefore since they philosophized in order to escape from ignorance, it is evident that they were pursuing knowledge in order to know, and not for any utilitarian end. And this is confirmed by the facts; for it was when almost all the necessities of life and the things that make for comfort and recreation had been secured, that such knowledge began to be sought.[1]

Plato and Aristotle held that the pursuit of knowledge was an end in itself. It was part of the good life for two reasons: first, what marks man out from the animals is the possession of reason, and so the cultivation of that faculty is essential for true happiness and for true excellence; and secondly, what the study of nature reveals is the beauty and order of the universe, contemplation of which helps a man to develop an orderly and noble character in himself. Yet although this theme is taken

[1] Based on the Oxford translation, *The Works of Aristotle translated into English*, edited by W. D. Ross (Oxford, Clarendon Press), *Metaphysics*, W. D. Ross (Vol. VIII, 2nd ed., 1928).

up by other ancient writers, it would be wrong to suppose that the beliefs and attitudes of Plato and Aristotle were shared by all their contemporaries, or even by all those who embarked on the study of nature.

First, there is a good deal of evidence that many ordinary people (not themselves scientists, to be sure) valued the practical arts. One may recall the references to agriculture, ship-building and mining, as well as medicine, in the passage setting out the debts of civilised man to Prometheus in the *Prometheus Bound* (436 *ff*), and the mention of agriculture and navigation in the chorus in the *Antigone* (332 *ff*) that speaks of the wonders achieved by man. Much as Plato himself advocated the pursuit of knowledge for its own sake, he was well aware that for the mass of ordinary people what counted was the practical utility of an inquiry. We have seen (p. 67) that in the *Republic* (527d) when Glaucon is asked whether astronomy should be included in the education of the guardians, he says:

> I certainly agree. Skill in perceiving the seasons, months and years is useful not only to agriculture and navigation, but also just as much to the military art.

It is true that Socrates is made to reply to this by saying:

> I am amused that you seem to be afraid lest the many suppose you to be recommending useless studies.

But the interesting thing about that remark is that it would imply that 'the many' generally *did* consider first and foremost the practical utility of any study, including those we should consider scientific disciplines. And that this was a common view is borne out by Isocrates, for example, who refers in the *Busiris* (23) to two quite different justifications offered for the study of astronomy, arithmetic and geometry, namely first, that they are useful in various ways, and second, that they are conducive to the attainment of virtue.

Secondly, when we turn to writers who themselves

engaged in scientific research, one whole group for whom the ideal of the contemplative life was clearly not the main motive is the doctors. They investigated the constitution of the human body, and the causes and cures of diseases, partly, no doubt, for the sake of knowledge itself, to satisfy their curiosity or 'owing to their wonder', as Aristotle put it. But the theories they adopted on those subjects were potentially of much more than merely academic interest to the patients of whom they had charge. The ultimate aim not only of the *Epidemics* and the surgical treatises, but also of many of the more speculative works on diseases and dietetics, was to convey information and to suggest ideas which would be of practical use to the doctor faced with the day-to-day problems of diagnosis, prognosis and treatment. As *On Ancient Medicine* notes (Chapter 1), those who based what they had to say concerning medicine on unfounded assumptions:

> are especially deserving of blame as their error relates to what is an art (*techne*), which all men use on the most important occasions and which they honour especially in the person of those who are good craftsmen and practitioners in it. For some medical practitioners are bad, others far superior: this would not be the case if there were no such thing as medicine or if no researches or discoveries had been made in it, but all would be equally inexperienced and ignorant in it, and the treatment of the sick would be entirely a matter of chance.

Medicine, as he says in Chapter 3, is an art founded on research and pursued for the sake of the health, preservation and nourishment of man.

Medicine was far from being the only one of the 'useful' arts on which technical monographs were composed in the fifth and fourth centuries. We hear of treatises written on agriculture (where the tradition of didactic literature goes back to Hesiod's *Works and Days*) and architecture especially, as well as on *mechanike*—the

134

study of mechanical devices – itself.[1] Thus in Xenophon's *Memorabilia* (IV, 2, 8 *ff*) Socrates implied that the sophist Euthydemus had in his possession a large number of treatises that dealt with medicine and architecture among such other subjects as mathematics, astronomy and rhetoric.[2] Democritus, as we have seen (p. 48), wrote on a variety of technical topics, including agriculture, painting and warfare. And in the *Politics* (1258 b 39 *ff*) Aristotle referred to the works that had been written on different aspects of *chrematistike*—the art of money-making, comprising agriculture, commerce and industry, and including such subjects as forestry and mining—and he cited treatises dealing with the various branches of agriculture in particular.

Finally, the author of the treatise *On Mechanics*—a follower of Aristotle, rather than Aristotle himself, although the work is found in the Aristotelian Corpus —also takes up the theme of the usefulness of knowledge. The bulk of the treatise consists of an account of the four simple machines, lever, pulley, wedge and windlass, together with brief discussions of specific problems connected, for example, with balances, the sailing of ships, the drawing of water from wells, even the extraction of teeth. Throughout the work the writer is chiefly interested in the mathematical principles involved in the devices he discusses. His aim is to give a theoretical, geometrical explanation of the phenomena, as when he suggests that the operation of a lever may be accounted for by the properties of a circle, and unlike some later mechanical writers, such as Ctesibius of Alexandria or Archimedes of Syracuse (both third century B.C.), he does not appear himself to be an inventor or an engineer. Yet while he responds to the element of

[1] Archytas is reputed to have been the first to apply mathematical principles to the study of mechanical devices.
[2] The sophist Hippias not only claimed to be able to teach all the arts, but also practised some of them. In the *Hippias Minor* (368a *ff*) we are told that he appeared on one occasion at Olympia dressed entirely in clothes that he had made himself.

the marvellous in mechanics, he also recognises its usefulness. 'In many things,' he says (847 a 13 *ff*), 'nature
acts contrary to our needs. . . . When therefore we have
to do something contrary to nature, this perplexes us
because of its difficulty, and we have need of art. And so
we call the branch of art that helps us in such perplexi-
ties mechanics,' and he goes on to quote with approval
the verse of the poet Antiphon: 'by art we conquer
where by nature we are overcome'.

Those who have written on ancient science have often
argued that an important difference between it and
modern science is that the ancients aimed merely to
understand nature and were not interested in trying to
use or control it. While true as a broad generalisation,
this tends to ignore the differences between the various
types of writers who engaged in the study of nature in
antiquity. Where the philosophers' ideal was the life
of leisure spent in 'contemplation', many of the early
Greek doctors were proud of practising an art, *techne*.
The usefulness of knowledge is often referred to, and
on some occasions it is clear that what the writer has in
mind is not that knowledge is conducive to happiness
or moral virtue, but that it has practical applications.
Nevertheless the dominant ideology among those who
investigated nature was that of the life of pure research,
and in view of the positive preference for non-utilitarian
over utilitarian studies expressed by such writers as
Plato and Aristotle, it is hardly surprising that the
Greeks were often slow to consider, or entirely failed to
consider, whether their theoretical knowledge could be
put to practical use. The view that we find in Francis
Bacon, that the goal of the acquisition of knowledge *is*
its practical benefits, is quite foreign to the ancient
world. Typically the writer of *On Mechanics* speaks of
recourse being had to 'art' only when nature itself does
not supply man's needs, and he does not envisage any
systematic exploration of the applications of mechanical
science to technology.

Some of the general motives underlying the 'inquiry

concerning nature' have been distinguished. But that term is much wider than our 'natural science' and we must recognise that the aims, expectations and motives of some ancient investigators of nature included a good deal that we should tend to dismiss as extraneous to science. An example that illustrates this clearly is the study of the heavenly bodies. Knowledge of the stars was sought both for its own sake (the 'philosophical' motive) and also in order to regulate the calendar (a practical one). But a third motive comes to the fore from about the middle of the fourth century on. Some of those who studied the stars did so in the belief that they influenced human destiny. This is not to say that the ancients drew no distinction between what we call astrology and astronomy. On the contrary they did so explicitly, at least in late antiquity, where we find Ptolemy for instance doing so in the opening chapters of the *Tetrabiblos*. Yet the distinction he drew is not one between a scientific and a non-scientific, or pseudo-scientific, study of the heavenly bodies, but between an exact, and a merely conjectural, branch of knowledge.

Although there is some evidence that astrological lore had begun to penetrate into the Greek world from Babylonia in the fourth century, it was not until after our period that it became an important reason for studying the heavenly bodies, and in particular for calculating the positions of the planets. But the fifth century already provides good examples to illustrate the complexity of the motives that sometimes underlie the investigation of nature. Mathematics and religion are inextricably interwoven in the researches of the Pythagoreans. The doctrine that all things are numbers provided the stimulus both for the empirical investigations that they carried out in acoustics, and for much fanciful speculation concerning the supposed similarities between things and numbers. Empedocles too was a religious teacher as well as a physicist. The relation between his poem *On Nature* and the *Purifications* is, admittedly, obscure, but it is striking that the same person who developed the element theory in the former claimed in the latter (fragment 112) to be an immortal god, no less:

> I, an immortal god, no longer a mortal, go about among you all, honoured as is fitting, crowned with fillets and flowery garlands.

And he goes on to describe how his fellow-citizens throng round him in their thousands, 'asking the path to gain, some desiring oracles, while others seek to hear the word of healing for all kinds of diseases'. The evidence we considered (pp. 54 f) from the treatise *On the Sacred Disease* suggests how some Greeks would have reacted to Empedocles' claim: the Hippocratic author submits the boasts of the 'magicians', 'purifiers' and sellers of charms or incantations to a devastating critique. Yet Empedocles' fragment shows that not all the 'purifiers' were mindless impostors: they included at least one philosopher who made a valuable contribution to physical theory.

There was, we said, no universally accepted view of the nature of scientific inquiry among ancient writers: equally it was not clear to the ancients themselves which of their investigations were leading down blind alleys, or were indeed in principle misconceived, and which were not. The dividing line between what could be considered a branch of knowledge and what was arbitrary speculation was drawn differently and with different degrees of clarity by different authors, and this was just as much a matter of disagreement between them as more specific issues concerning, for example, the relevance of final causes—on which Plato and Aristotle opposed Empedocles and the atomists—or the value of observation—on which Plato and Aristotle themselves took opposite sides. We may conclude that curiosity and the desire for knowledge, both theoretical and practical, were the most general motive forces behind the investigation of nature in antiquity, but we must acknowledge that the investigations that those motives prompted took very different forms, reflecting differences not merely in the scientific interests, but also in the philosophical and religious beliefs, of the authors concerned.

Given the differences in the way in which the 'inquiry concerning nature' was construed, how far is it possible, nevertheless, to summarise the general character of science in our period? The distinguishing features of early Greek science are often thought to lie less in its content than in its methods. In particular it is often argued that what marks out ancient science from science since the Renaissance is that the ancients failed to appreciate the value and importance of experimentation. This too is a propositon which, while true in the main, needs some qualification.

The first reservation to make is that experiment is simply not relevant to many of the problems that the Greek scientists were interested in. Astronomy and meteorology figure prominently in the speculations of the earliest natural philosophers, but in neither field was it open to them to employ direct experimentation. This is obvious in the case of astronomy, since the movements of the heavenly bodies cannot be controlled—although the regularity with which those movements repeat themselves provides an alternative means of checking hypotheses. But the Greeks had no means of investigating such phenomena as lightning and thunder directly either. The most they could do was to argue by analogy from more familiar phenomena, and from Anaximander onwards this was indeed their stock method of tackling these problems.

Again the experimental method was only of very limited usefulness on the fundamental problem of physics, the question of the ultimate constituents of matter. Although quite simple experiments would have yielded useful information about the nature of certain compounds, the principal controversy between atomism and the qualitative theory of Aristotle, for example, was not one that could be settled by an appeal to either observations or experiments, since the controversy

139

turned on the question of the type of account that was to be attempted.

In such fields as these, then, it hardly makes sense to talk of the Greeks failing to use the experimental method, since it was either impracticable or quite impossible to devise experiments that would resolve the issues in question. This does not, however, apply to many other problems in physics and biology. But then the second reservation that needs to be made concerns the extent to which the Greeks failed to bring their theories to the test. That they never experimented is patently untrue. The most thorough, systematic and fruitful experiments in Greek science date, as one would expect, from much later than the period we have been considering, for example Ptolemy's experiments in optics or those of Galen on the nervous system (both in the second century A.D.). But even in the fifth and fourth centuries B.C. experimentation was used on a limited scale. We have reviewed (pp. 30 f) the evidence for Pythagorean experiments in acoustics: although much of this evidence is untrustworthy, some is all the more impressive as it derives from a source quite hostile to the use of such methods, namely Plato.

Several simple experiments are also referred to in the Hippocratic writers and in Aristotle. It is certainly true that the results of these tests are sometimes misreported —or the tests themselves were not carried out correctly —and more often they do not demonstrate what their authors thought they did. A typical example of this is the experiment described in the treatise *On Airs, Waters, Places,* Chapter 8, where the writer suggests leaving a bowl of water out of doors on a cold night to freeze, and says that when the water is thawed it is found on being remeasured to be less than the original quantity. If there was any loss of water, this was, no doubt, due to evaporation and not an effect of the freezing. But what is more striking is that the writer cites this experiment as support for his theory that when water freezes, the 'bright, light and sweet' part is separated out

140

from the 'muddy and heavy'. He asserts that the experiment shows that freezing causes the 'lightest and finest' part of the water to 'be dried up and disappear', while the heavy and coarse part is left behind. Nevertheless there are examples of simple experiments that were successfully brought to bear on certain problems. Thus in his discussion of why the sea is salt, Aristotle states the view that it is becoming saltier: it does not, he claims, lose any appreciable quantity of salt through evaporation and he refers to an experiment in this connection. 'We can assert on the basis of experiment (*pepeiramenoi*),' he says in the *Meteorology* (358 b 16 *ff*), 'that salt water when evaporated forms fresh and the vapour does not form sea water when it condenses again.'

In some instances early Greek scientists did follow up specific physical and biological theories by conducting simple experiments. Yet the occasions on which they did so were limited, and it is not difficult—with the benefit of hindsight—to suggest examples where the Greeks failed to carry out experiments that were both open to them and relevant to problems they discussed. Thus simple experiments such as Redi was to perform in the seventeenth century might have been carried out to establish whether the worms found in decaying meat were spontaneously generated by the putrefaction of the meat itself or were directly derived from fly droppings, and such experiments would have shown the Greeks that some of their common assumptions concerning the spontaneous generation of animals were incorrect. It is, however, obvious that the idea of conducting such tests would only occur when the doctrine of spontaneous generation as a whole had been called in question.

Then a more important point is that such experiments as were performed by the Greeks were usually carried out with the set purpose of supporting the writer's own theory. The appeal to experiment was an extension of the more usual notion of appealing to evidence: experimentation was a corroborative, far more than a heuristic, technique. Tests were conducted to confirm the desired result, and it is only in late antiquity that we find examples where attempts were made to vary the conditions of experiments systematically in order to isolate causal relations.

We have found plenty of evidence, in the course of this study, for the use of empirical methods in early Greek science. While experimentation was still quite rare, the Greeks produced many skilled and persistent observers, both in biology and medicine, and in astronomy. Nevertheless the impression that much of the history of early Greek science leaves is one of the dominant role of abstract argument. This was partly due to the nature of the problems examined. To investigate the problem of change, the question of the foundations of knowledge had to be discussed. If much of ancient science is closely linked to philosophy, there are good reasons for this in the need to tackle certain fundamental epistemological issues as a preliminary to embarking on the scientific questions themselves. In their attempts to clarify the nature of the problems, the Greek scientists—both philosophers and medical writers—repeatedly refer to and criticise one another's ideas. To have instituted this tradition of rational debate was, we suggested, one of the essential contributions of the Milesians, and thereafter the rivalry between individuals and between groups undoubtedly did much to stimulate interest and discussion not only on specific scientific problems, but also on second-order issues concerning the nature of the inquiry and the methods proper to it.

But if keenness in dialectic was one of the strengths of early Greek science, it was also a source of weakness. The discussion of scientific problems was often marred by being conducted too much like an argument in a court of law. As the treatise *On the Nature of Man* shows, even such obscure questions as the constitution of the human body were debated before public audiences in the late fifth century (see above, pp. 11 and 61). Such debates played, we may suppose, a very necessary part in the dissemination of ideas: although the fifth and fourth centuries saw an increase in the circulation of written texts—one of the first important libraries was Aristotle's—books were still rare objects even at the end of our period, and the role of personal contact and conversation in the transmission of knowledge was correspondingly much greater than in modern society. But in the context of the public debates referred to in *On the Nature of Man* there was a tendency for the speaker to present only one side of the case, to state his own view and undermine his opponents' arguments, but to leave to another speaker the exploration of the weaknesses of his own position. Moreover, what counted with the audience who adjudicated such contests was the rhetorical skill of the speakers, rather than their knowledge of the scientific subject being debated. Of course the problems of science were not always discussed in such unfavourable circumstances. The relation between Aristotle and his colleagues in the Lyceum was, no doubt, very different from that between the disputants in the contests described in *On the Nature of Man*. Yet it remains the case that the Greeks not only enjoyed, and excelled in, debate, but also were often too ready to see in *argument* the solution to their problems. At the same time this very criticism is one of which the Greeks themselves were at least to some extent aware. We found (pp. 59 *ff*) the author of *On Ancient Medicine*, for example, castigating those who spoke on scientific subjects in such a way that neither the speaker himself nor his audience were sure 'whether what was said was true or not, since there is no criterion to which one should refer to obtain clear knowledge'.

Throughout this discussion I have insisted on the complexity of the data we are dealing with when we speak of 'Greek science'. Yet an attempt must be made, in conclusion, to take stock of what the first 250 or 300 years of Greek science may be said to have achieved. To begin with, we must observe that even at Athens, and even at the end of the period we have been considering

—after the founding of the Lyceum—science was an interest of only a handful of individuals, and elsewhere scientists were rare indeed. There were, as we saw, no financial incentives to become a scientist, and science as such received no state support. The idea that dominates our own society, that science holds the key to material progress, was quite foreign to the ancient world, and without that idea the scale of scientific activity was and always remained minute by modern standards. All the scientists we have been discussing in this study would be outnumbered by the scientific teaching staff of a medium-sized modern university.

Yet the achievements of the few isolated individuals who were interested in different aspects of the 'inquiry concerning nature' were far from negligible. First, and most obviously, there were advances in factual knowledge in various fields. This is particularly true of the descriptive branches of biology, especially anatomy and zoology, thanks largely, if not entirely, to the work of Aristotle. In astronomy also by the end of the fourth century the Greeks were beginning to prove themselves capable observers, as we see, for instance, in the increased accuracy of measurement of the lengths of the seasons.

Second, and more important, was the progress made in grasping the nature of certain problems. The chief example of this is the problem of change. We saw how an awareness of this problem develops among the earliest Presocratics and how Parmenides' denial of the possibility of change set the main problem that later physicists had to tackle. Thereafter Aristotle's explicit distinction between matter and moving cause enabled the two questions, (i) of the ultimate constituents of material objects, and (ii) of the efficient causes of different types of change, to be kept separate. Whereas at the beginning Thales was groping towards some conception of an originative substance, by the mid fourth century the Greeks had developed a rich vocabulary of terms, such as 'substance', 'quality', 'matter', 'substratum', 'cause' and so on, in which to discuss the cluster of problems related to change.

In more specific fields of inquiry, too, the problems came to be defined more clearly, as when Plato formulated what was for long to remain the main problem in astronomy, that of reducing the irregular courses of the planets to regular motions, or when Aristotle set out the issues on the biological problems of reproduction and heredity. Even when the concrete theories that the Greeks put forward contain little of lasting value, later science often owes to the early natural philosophers the first clear statement of some of the fundamental problems.

But the third and perhaps most important achievement of our period was the development of the two key methodological principles, (i) the application of mathematics to the understanding of natural phenomena, and (ii) the notion of undertaking empirical research. The first idea owes most to the Pythagoreans and to Plato, and there was one notable example of it in practice in Eudoxus' astronomical model. The idea of undertaking deliberate research was a natural extension of the curiosity that motivated all Greek scientists to a greater or lesser extent, but first the Hippocratics, and then Aristotle, especially, showed what it was to set about collecting detailed information in a particular field of inquiry. Aristotle's defence of his method in connection with the study of animals testifies both to the resistance that this technique met from some quarters and to Aristotle's own appreciation of its value.

Today we take both these methodological principles so very much for granted that it demands an effort of the imagination to see that they required discovery. Yet this must be acknowledged to be the case. Indeed they needed to be not merely discovered by the Greeks, but also rediscovered in the Renaissance. The role of Plato as the chief ancient advocate of the mathematisation of the sciences was in fact recognised by such men as Galileo and Kepler. As for the second idea, it is clear that while Vesalius, for instance, rejected many of the views which he found in Aristotle and Galen, he can be represented as marking a return to their empirical methods after a long period in which the practice of dissection, indeed the practice of observation itself, had fallen into disuse.

Obvious as these two ideas seem to us, then, it is they, rather than any of the concrete theories of our period, even such brilliant constructions as Eudoxus' astronomy or Aristotle's comprehensive natural philosophy, that represent, in the long run, the most important legacy of the early Greeks to later science. The question of the comparative successes and failures of the later Greeks themselves in putting these methodological principles to work is the second part of the story of Greek science.

Select Bibliography

I SOURCES: TEXTS AND TRANSLATIONS

A   General
A Source Book in Greek Science, edited by M. R. Cohen and I. E. Drabkin (second edition, Cambridge, Mass., Harvard University Press, 1958). This does not include cosmology, but otherwise provides a full selection of the most important passages in translation with a useful bibliography.

B   The Presocratic Philosophers
TEXT: The chief source book is *Die Fragmente der Vorsokratiker,* 3 vols, edited by H. Diels and W. Kranz (6th edition, Berlin, Weidmannsche Verlagsbuchhandlung, 1951–2).
TRANSLATIONS: An English translation of the fragments in Diels-Kranz is available in K. Freeman, *Ancilla to the Pre-Socratic Philosophers* (Oxford, Blackwell, 1948; Cambridge, Mass., Harvard University Press), but this should be used in conjunction with a commentary; G. S. Kirk and J. E. Raven, *The Presocratic Philosophers* (Cambridge, University Press, 1957) both translates and discusses the most important texts; W. K. C. Guthrie, *A History of Greek Philosophy,* vols. 1 and 2 (Cambridge, University Press, 1962, 1965) is more extensive and contains a full bibliography.

C   The Hippocratic Corpus
TEXT: The most recent complete edition is that of E. Littré, *Œuvres complètes d'Hippocrate,* 10 vols (Paris, J. B. Baillière, 1839–61).
TRANSLATIONS: The best translation of the most important treatises is J. Chadwick and W. N. Mann, *The Medical Works of Hippocrates* (Oxford, Blackwell, 1950), but this selection can be supplemented by the four-volume Loeb Edition of Hippocrates by W. H. S. Jones and E. T. Withington (Cambridge, Mass., Harvard University Press; London, Heinemann, 1923–31).

D Plato

TEXT: The standard text is that of J. Burnet, *Platonis Opera,*
5 vols. (Oxford, Clarendon Press, 1899–1906).
TRANSLATIONS: The best complete translation is that of
the Loeb edition. The most important dialogue for the history of Greek science is the *Timaeus,* on which there are two
excellent commentaries: A. E. Taylor, *A Commentary on
Plato's Timaeus* (Oxford, Clarendon Press, 1928) and F. M.
Cornford, *Plato's Cosmology* (London, Routledge & Kegan
Paul, 1937; New York, Humanities Press). Cornford's translation of the *Republic* is also valuable: *The Republic of Plato*
(Oxford, Clarendon Press, 1941; New York, Oxford University Press).

E Mathematicians and astronomers

TEXTS: Eudoxus: F. Lasserre, *Die Fragmente des Eudoxos
von Knidos* (Berlin, De Gruyter, 1966). Heraclides: F.
Wehrli, *Herakleides Pontikos* (Die Schule des Aristoteles,
vol. VII) (2nd edition, Basle, Benno Schwabe, 1969).
TRANSLATIONS: Most of the important passages are translated in either T. L. Heath, *Greek Astronomy* (London,
Dent, 1932) or the same author's *A History of Greek Mathematics,* vol. I (Oxford, Clarendon Press, 1921; New York,
Oxford University Press).

F Aristotle

TEXT: The most recent complete edition is that of the Berlin
Academy (Berlin, Reimer, 1831–70), but many of the treatises
have now been edited in the Oxford Classical Texts series.
TRANSLATIONS: The best complete translation is *The
Works of Aristotle translated into English,* edited by W. D.
Ross, 12 vols. (Oxford, Clarendon Press, 1908–52), but many
of the Loeb translations are also excellent, notably A. L.
Peck's *Parts of Animals* (Cambridge, Mass., Harvard Uni-
versity Press; London, Heinemann, 1937), *Generation of
Animals* (1943) and *Historia Animalium* (vol. I, 1965), and
W. K. C. Guthrie's *On the Heavens* (Cambridge, Mass.,
Harvard University Press, 1939).

SELECT BIBLIOGRAPHY

II SECONDARY READING

A *General*. The most important works in English are:

S. Sambursky, *The Physical World of the Greeks* (trans. M. Dagut, London, Routledge & Kegan Paul, 1956; New York, Humanities Press, Collier-Macmillan (paper) 1956)

O. Neugebauer, *The Exact Sciences in Antiquity* (second edition, Providence, R.I., Brown University Press, 1957)

M. Clagett, *Greek Science in Antiquity* (London, Abelard-Schuman, 1957; New York, Collier-Macmillan (paper))

B. Farrington, *Greek Science* (revised one vol. edition, London, Penguin Books, 1961; Baltimore, Md., Penguin Books)

Two older works are also still worth consulting:

A. Reymond, *History of the Sciences in Greco-Roman Antiquity* (trans. R. G. de Bray, London, Methuen, 1927; New York, Biblo & Tannen)

W. A. Heidel, *The Heroic Age of Science* (Baltimore, Carnegie Institution of Washington, 1933)

There are also useful sections on Greek science in:

G. Sarton, *A History of Science*, vol. 1 (London, Oxford University Press, 1953; New York, W. W. Norton & Company, Inc., 1970)

J. Needham, *A History of Embryology* (2nd edition, Cambridge, University Press, 1959)

M. Hesse, *Forces and Fields, The concept of action at a distance in the history of physics* (London, Nelson, 1959; Totowa, N. J. Littlefield, Adams Co., 1961)

B *Background and comparative material*

V. Gordon Childe, *Man Makes Himself* (London, Watts, 1936; New York, New American Library, 1952), *What Happened in History* (London, Penguin Books, 1942; Baltimore, Md., Penguin Books), *The Prehistory of European Society* (London, Penguin Books, 1958)

H. Frankfurt (ed.) *Before Philosophy* (London, Penguin Books,1949; Baltimore, Md., Penguin Books)

C. Singer, E. J. Holmyard, A. R. Hall (ed.), *A History of Technology*, vols. 1 and 2 (Oxford, Clarendon Press, 1954, 1956)

R. J. Forbes, *Studies in Ancient Technology*. 9 vols, have so far appeared, some in a second edition (Leiden, Brill, 1955, in progress)

The most important of the works of C. Lévi-Strauss dealing with 'Primitive thought' is *The Savage Mind* (London, Weidenfeld & Nicolson, 1966; University of Chicago Press, 1967)

C *Presocratics* (further bibliography in Guthrie, 1 *B* above)

D. M. Balme, 'Greek Science and Mechanism', *Classical Quarterly*, no. 33 (1939), pp. 129–38, and no. 35 (1941), pp. 23–8

E. Schrödinger, *Nature and the Greeks* (Cambridge, University Press, 1954)

K. R. Popper, 'Back to the Presocratics', *Proceedings of the Aristotelian Society*, no. 59 (1958–9), pp. 1–24, reprinted in *Conjectures and Refutations* (2nd edition, London, Routledge & Kegan Paul, 1965; New York, Harper & Row, Torch Book, 1968) pp. 136–53

G. S. Kirk, 'Popper on Science and the Presocratics', *Mind*, no. 69 (1960), pp. 318–39

G. E. R. Lloyd, 'Popper versus Kirk: a controversy in the interpretation of Greek Science', *British Journal for the Philosophy of Science*, no. 18 (1967), pp. 21–38

D *Hippocratics*

W. A. Heidel, *Hippocratic Medicine: its spirit and method* (New York, Columbia University Press, 1941)

W. H. S. Jones, *Philosophy and Medicine in Ancient Greece* (Suppl. 8 to the Bulletin of the History of Medicine, Baltimore, Johns Hopkins Press, 1946)

L. Edelstein, *Ancient Medicine* (Baltimore, Johns Hopkins Press, 1967)

E *Plato*

P. Shorey, 'Platonism and the History of Science', *Proceedings of the American Philosophical Society*, no. 66 (1927), pp. 159–82

G. C. Field, 'Plato and Natural Science', *Philosophy*, no. 8 (1933), pp. 131–41

SELECT BIBLIOGRAPHY

I. M. Crombie, *An Examination of Plato's Doctrines*, vol. 2 (London, Routledge & Kegan Paul, 1963; New York, Humanities Press, 1963)

G. E. R. Lloyd, 'Plato as a Natural Scientist', *Journal of Hellenic Studies*, no. 88 (1968), pp. 78–92 (contains further bibliographical references)

F   Mathematics and astronomy

T. L. Heath, *Aristarchus of Samos* (Oxford, Clarendon Press, 1913; New York, Oxford University Press)

O. Neugebauer, 'The History of Ancient Astronomy; Problems and Methods', *Journal of Near Eastern Studies*, no. 4 (1945), pp. 1–38

B. L. van der Waerden, *Science Awakening* (trans. A. Dresden, Groningen, Noordhoff, 1954; New York, Oxford University Press, 1961; paperback, John Wiley & Sons, Inc.)

F. Lasserre, *The Birth of Mathematics in the Age of Plato* (trans. H. Mortimer, London, Hutchinson, 1964)

G   Aristotle

T. E. Lones, *Aristotle's Researches in Natural Science* (London, West, Newman & Co., 1912)

I. E. Drabkin, 'Notes on the Laws of Motion in Aristotle', *American Journal of Philology*, no. 59 (1938), pp. 60–84

D'A. W. Thompson, 'Aristotle the Naturalist' in *Science and the Classics* (London, Oxford University Press, 1940), pp. 37–78

R. McKeon, 'Aristotle's Conception of the Development and the Nature of Scientific Method', *Journal of the History of Ideas*, no. 8 (1947) pp. 3–44, reprinted in *Roots of Scientific Thought* (ed. P. P. Wiener and A. Noland, New York, Basic Books, 1957) pp. 73–89

F. Solmsen, *Aristotle's System of the Physical World* (New York, Cornell University Press, 1960)

H   Methodology of ancient science

H. Gomperz, 'Problems and Methods of Early Greek Science', *Journal of the History of Ideas*, no. 4 (1943), pp. 161–76, reprinted in *Roots of Scientific Thought* (ed. P. P. Wiener and A. Noland, New York, Basic Books, 1957) pp. 23–38

L. Edelstein, 'Recent Trends in the Interpretation of Ancient Science', *Journal of the History of Ideas*, no. 13 (1952), pp. 573–604, reprinted in *Roots of Scientific Thought*, pp. 90–121

G. E. R. Lloyd, 'Experiment in Early Greek Philosophy and Medicine', *Proceedings of the Cambridge Philological Society*, no. 190 (n. s. 10) (1964), pp. 50–72

Index

INDEX

craftsman (*dēmiourgos*), 3
51, 106, 126, 134; in
Plato, 70, 72–3
critical days, 58
Croton, 24, 51
Ctesibius of Alexandria,
135
cube, problem of duplica-
tion of, 34
Cyrene, 51

Dalton, 45
debate, 10–15, 51, 58, 61,
142–3
dēmiourgos, see craftsman
Democedes, 53
Democritus of Abdera, 39,
45–9, 63, 77, 112, 135; see
also atomism
demonstration (*apodeixis*),
99–100
design, 72, 78, 98, 120–1,
124
diagnosis, 52, 55–9, 134
diet, dietetics, 50, 52, 134
Diogenes of Apollonia, 73
Diogenes Laertius, 126
dissection, 64, 104–5, 116,
146
doctors, ch 5, 125, 127, 130,
134
doxographers, 10, 80–1
dynamics, 109–10, 112–15,
123

earth, 9, 11, 16, 17, 20, 28,
30, 39–42, 61, 74–7, 80,
82, 83, 86, 94–5, 107–10
earthquakes, 9, 16
eccentric circles, theory of,
91, 94–6
eclipses, 7–8, 29, 81
ecliptic, 81, 83, 85–6
education, 50–1, 66, 67–9,
127–8
Egyptians, 2, 4–6, 11–12,
13, 109, 132
Eleatic philosophers, 39,
46
elements, 19, 39–41, 44–6,
59–61, 64, 74–6, 106–12
embryology, 63–4, 118–20
154

Empedocles of Acragas,
10, 11, 38, 39–46, 49, 73–
74, 76, 120–1, 126, 130–1,
137–8
  On Nature, 137–8
  Purifications, 42, 137–8
endoxa (common opinions),
102
epicycles, 91, 94–7
epiktētos (acquired), 119
epilepsy, 54–5
epistēmē (knowledge), xv, 99
epistemology, see know-
ledge
Erasistratus, 127
erga (facts), 103
ethics, 27, 42, 66, 72, 79,
104, 121–2
Euclid, 31, 35
Euctemon, 82, 90, 92
Eudemus, 123
Eudoxus of Cnidus, 82,
86–94, 96–8, 145–6
Euripides, 131
Euthydemus, 135
evolution, 18, 121
experiment, 3, 30–1, 115,
139–42

Farrington, B., 9
fees (of doctors and sophists), 53, 127–8
fevers, classification of, 58
final cause, 105–7, 115,
118, 120–2, 138
fire, 17, 28, 39–42, 61, 74,
77, 107–10, 122
form, 25; in Aristotle,
105–7, 115, 118, 120–1; in
Plato, 67, 70–2, 124, 131

Galen of Pergamum, 127,
140, 146
galeoi (dog-fish), 18, 118
Galileo, 145
geology, 49, 60
geometry, 31–4, 67–9, 74–7,
84, 90, 96–8, 100, 123,
128, 133, 135
gnomon, 97
gods, 9, 16, 19, 54, 122,
138; craftsmen-gods in
Plato, 73

growth, problem of, 44,
62–3, 118
gynaecology, 50

happiness, 71, 131–2, 136
harmoniā (attunement), 27,
as synonym for *Philiā*
(Love) in Empedocles, 42
heart, 72, 116, 117
Heath, T. L., 34, 89
Heraclides Ponticus, 94–
97
Heraclitus of Ephesus, 10,
36–7, 43, 126
heredity, 63, 119–20, 145
Herodotus, 14, 53, 130
Herophilus, 127
Hesiod, 9, 10, 19, 40, 81,
134
Hestia, 28
Hipparchus of Nicaea, 97
Hippias of Elis, 50, 127,
128
Hippocrates of Chios, 34
Hippocrates of Cos, 34, 50
Hippocratic writers, 5, 11,
ch 5, 73, 119, 134, 138,
140–5
  Epidemics, 53, 56–8, 134
  On Airs, Waters, Places,
  50, 140–1
  On Ancient Medicine, 11,
  51, 58–61, 62, 65, 134,
  143
  On Breaths, 58
  On Diseases IV, 63
  On Fractures, 64
  On Generation, 63
  On Joints, 64
  *On Regimen in Acute Dis-
  eases*, 59
  On the Heart, 64
  On the Nature of Man, 11,
  61–2, 142–3
  On the Nature of the Child,
  62
  On the Sacred Disease, 51,
  54–5, 138
  Precepts, 53
  Prognostic, 52–3, 55–6
Hippolytus, 17, 20
historiā (inquiry), xv, 131
Homer, 9
humours, 57, 61

INDEX